THE
SWORD
and
THE PEN

THE
SWORD
and
THE PEN

SELECTIONS FROM
THE WORLD'S GREATEST
MILITARY WRITINGS

PREPARED BY

Sir Basil Liddell Hart

EDITED BY

Adrian Liddell Hart

A Martin Dale Book

THOMAS Y. CROWELL COMPANY
NEW YORK/*Established 1834*

To the memory of my father

CONTENTS

I

The Ancient World

II

Renaissance and Reformation

III

The Age of Reason

IV

The Revolution

V

The Later Nineteenth Century

VI

The Twentieth Century

Men grow tired of sleep,
love, singing and dancing,
sooner than of war.
—HOMER

THE
SWORD
and
THE PEN

INTRODUCTION

> "War is too serious a thing
> to be left to military men."
> —TALLEYRAND

In 1968 my father agreed to prepare an anthology on war: "a selection of the great writing on military topics by Historians and Authorities of all nations from ancient times to the present day." At that time he was working on his *History of the Second World War* which he finished just before his death in 1970. At the publisher's request I took over this project.

I have made use not only of my father's proposed outline, indicating those writers whom he was considering for inclusion, but also of his military library as a whole—now in the possession of King's College, London University. This contains the marginal notes that he had made on many of the military classics, as well as the correspondence with historians and military men which he had conducted over fifty years on the theory and practice of war.

I have not, however, followed an academic pattern, but rather sought to preserve an open—if sometimes indirect—approach to the meaning of war. The project has been an exploration not only into war but into the minds and nature of those who have engaged in it, with sword and pen, in time past. . . .

"My mind was scarce opened when my father gave me the first lesson of tactic," wrote Guibert, the eighteenth-century pioneer of Napoleonic warfare. "On returning home we would resume our game. . . . We next formed two armies and each took command of one of them. Then in different types of country, represented at chance by the arrangement of the pasteboard plans, we made our armies manoeuvre; we made them execute marches; we made choice of positions; we formed in battle one against the other. We afterwards reasoned out between ourselves what we had done. My father encouraged my questions and even contrary opinions. The nights frequently passed in this occupation, so much did this study absorb us, so well did my teacher know how to make it interesting."

Some of my own earliest memories are of war games played in the garden on summer afternoons, like Sterne's Uncle Toby and Corporal Trim in *Tristram Shandy,* or in my father's study of an evening.

I accompanied my father, too, on military maneuvers at home and abroad, which he was covering as a military correspondent for the *Daily Telegraph* and the *London Times.* The exercises were often confusing; the battles sometimes seemed to go on a long time, especially when it rained, as it often did. On one occasion I found myself on the battlefield of Waterloo, like Stendhal's Fabrizio. However, I did not succeed in attaching myself to Marshal Ney, the hero of the Moskowa, but to King Leopold of the Belgians who was watching his army maneuver from the top of the convenient Victory Monument. Such are the revolutions of fortune that I fell into a manure heap at the farm of La Haye Sainte on the same day—while the King was later to be damned by responsible statesmen and popular writers as one who had, even at this time, been plotting to destroy his allies.

Where do war games end? "For several years the bell of my flat would ring on Christmas Eve," relates von Kuhl,* the aide of Field Marshal von Schlieffen. "A courier would bring his Christmas present, a great military situation designed by him for the set task of working out an operational plan. He would have been very surprised if the solution had not been in his hands on the evening of Christmas day." Schlieffen's game may have meant a not so merry Christmas for his aide. For others the great plan, to which he devoted almost his entire life, was to mean death on the western front—though it would be unfair to put the whole blame on the former chief of the German general staff for what, anyhow, miscarried after his death.

"But war is no pastime," says Clausewitz. "It is a serious means for a serious object." I used to accompany my father to the house of Lloyd George, where he was assisting the statesman in preparing his war memoirs. While I sat silently in a corner, the great war leader went on talking about Haig and other British commanders in terms in which, at school, a teacher might refer—though considerably less eloquently—to particularly stupid and dishonest boys. Yet these were the men we were expected to revere, especially on Armistice Day.

When real war was supposed to come again, in 1939, I was in France. I recall the son of the Polish ambassador explaining to me on

* Gerhard Ritter, *The Schlieffen Plan.*

the beach one morning how the Polish cavalry would be in Berlin in a few days. I had learned enough to be sceptical. Before many months I was guarding a bridge at Totnes, in Devonshire, in the service of what were optimistically christened the Local Defense Volunteers. My father thought it was an important bridge and route. "It is worth recalling," he wrote discreetly in an article at the time, "that the last and only successful invasion of England in modern times was made at Torbay (in 1688). . . ."

Two hundred years before, the military writer, Henry Lloyd, dressed as a priest, had spied out the invasion possibilities of the area for Marshal Saxe. "There is but one narrow road," he wrote in *The Military Rhapsody,* "which goes from Dartmouth to Newton Bushell, near which the tide flows. A few miles from Dartmouth, a branch of it turns off to Totness, and several towns on the west, as Torbay, Paignton, etc., come into the main road, leading to Hall Down or to Plymouth."

Another year and I was playing other war games. By this time I was aware that such games were different when played out on a small ship's deck in a freezing ocean, into which one might at any moment be blown.

"Such a sweet felicity is that noble exercise," as Bacon expresses it, "that he that hath tasted it thoroughly is distasteful for all other. And no marvel; for if the hunter takes such solace in the chase; if the matches and wagers of sport pass away with such satisfaction and delight; if the looker-on be affected with pleasure in the representation of a feigned tragedy, think what contentment a man receiveth, when they, that are equal to him in nature, from the height of insolency and fury are brought to the condition of a chased prey; when a victory is obtained, whereof the victories of games are but counterfeits and shadows; and when, in a lively tragedy, a man's enemies are sacrificed before his eyes to his fortune."*

After a man's death some reassessment of his life's work usually takes place. This is not, however, the subject for an anthology. I have included two extracts from the works of my father—he had, in any event, intended to leave any such selection to a collaborator. As with others, I have chosen them, not because I happen to agree with what is said, but primarily because they seem to express well a view of war —and to represent the influence of the writer in his time.

* Francis Bacon, *Essays.*

It was through his detailed criticism of such contemporary battles as Passchendaele, addressed to a wide readership which shared, to some extent, his personal engagement, that he came to be seen as, in his way, a revolutionary—and in some quarters, as a prophet. Yet the strategy of indirect approach was essentially a traditional and, indeed, classical doctrine. Although the summarized extract that I have given is taken from one of his later works, elaborating on an earlier one, it should be seen in the context of a heritage, a view of war—and life—which goes back beyond the eighteenth-century values that he admired, to the Elizabethans and the Greeks.

He acknowledged more recent debts. If I have omitted from this anthology some of these, British and French particularly, which he himself was considering for inclusion, it is because I have judged it more appropriate to maintain a different perspective, and not because he would have wished to separate himself from those who had stimulated or helped him.

Sometimes the subject of war-writing becomes of incidental significance. Who is concerned at this stage about who won the Battle of Omdurman—or how? "They displayed the virtues of barbarism. They were brave and honest. The smallness of their intelligence excused the degradation of their habits. Yet their eulogy must be short for though their customs, language, and appearance vary, the history of all is a confused legend of strife and misery."*

Churchill lived long enough to see a native military government installed where he had warned that "the political supremacy of an army always leads to . . . the degradation of the peaceful inhabitants through oppression and want, to the ruin of commerce, the decay of learning, and the ultimate demoralization of the military order through overbearing pride and sensual indulgence." *The River War* is now of interest because it tells us something about the writer—and because it contributed to the development of Churchill's own reputation.

The Crimean War maintains a claim on our attention because Russell's *Despatches,* as the first war correspondent of the *London Times,* were instrumental in bringing about changes in public attitudes and the fall of the government responsible. And it was in the Crimean War, and in even obscurer campaigns, that Tolstoy derived the experience that was to be developed in *War and Peace,* bringing a

---

* Winston Churchill, *The River War.*

view of war to the attention of millions who had scarcely heard of Austerlitz, let alone Sebastopol.

"The mechanical part of war is insipid and tedious in description," Marshal Saxe wrote, "of which the great captains being sensible, they have studied to be rather agreeable than instructive in their writings upon the subject; the few books which treat of war as an art are but small in esteem. . . ." Saxe's own *Reveries* (which Carlyle described as a "strange military farrago dictated, as I should think, under the influence of opium") may be considered an exception.

"Great commanders," my father wrote, "have mostly been dull writers. Besides lacking literary skill in describing their actions, they have tended to be cloudy about the way their minds worked. In relating what they did, they have told posterity little about how and why." Napoleon may make us wonder how a man who commanded such awe from a generation of military historians could dictate so much which is banal.

In those works on war which treat of strategy and tactics, a broad division can be made between the intellectual, material, and moral aspects—though most writers have made the point, with varying emphasis, that these are interrelated. "Every theory," wrote Clausewitz, "becomes infinitely more difficult from the moment it touches on the province of moral quantities."

On the intellectual aspect, the science of war, one would scarcely look to novelists and other imaginative writers for expositions. Some authorities have, indeed, been inclined to the view that one should not look beyond the ranks of professional military instructors. Yet, despite the ever-increasing sophistication of military organization—or, perhaps, on account of it—fictional military expositions are no longer unfashionable in conditions that focus attention on subversive or irregular warfare. As St. Just once remarked, while rallying the Revolutionary armies, "*les malheureux sonta la puissance de la terre.*"

However, I am conscious that these selections, as a whole, may not do justice to the professional contribution, whether in the field or in the classroom. "The most eminent military thinkers," writes Professor Michael Howard, "sometimes do no more than codify and clarify conclusions which arise so naturally from the circumstances of the time that they occur simultaneously to those obscurer but more influential figures who write training manuals and teach in service colleges. And sometimes strategic doctrines may be widely held which cannot be attributed to any specific thinkers, but represent simply the consensus of opinion among a large number of professionals who had undergone a formative common experience."*

This is not the place, I think, to try (if that is possible) to determine the comparative influence of military commanders or writers on the conduct of war throughout the ages—or their particular influence on each other. Neither the inclusion of a writer nor the placing of an extract next to another should be treated in this sense. Writers—and commanders—acknowledge a debt for various reasons, or omit to do so. An extensive pattern can be traced in this respect, but whether it proves much more than that "great" thinkers and even "great" men of action, with such notable exceptions as Jenghiz Khan, are usually well read, is open to doubt. Military history, when superficially studied, will furnish arguments in support of any theory or opinion.

"In reply to your query with reference to the integration of the Inchon Campaign with Wolfe's Quebec operation," General MacArthur wrote to my father in 1959, "so much time has elapsed since then that I would hesitate to attempt a categorical reply. That I have read and studied your account in *Great Captains Unveiled*† is unquestionable. . . . The most indispensable attribute of the great Captain is imagination." Wolfe, we know, was a well read general; "I had it from Xenophon," he explained at Louisbourg—while Xenophon had discussed matters with Socrates, who gave him the benefit of his own military experience.

"There has been no illustrious captain who did not possess taste and a feeling for the heritage of the human mind," de Gaulle wrote.‡ "At the root of Alexander's victories one will always find Aristotle." What does this signify? That de Gaulle was an intellectual soldier who already saw himself as a great captain—almost certainly. That Alexander was stimulated by Aristotle's teaching, very likely. That Aristotle's teaching had any particular bearing on the battle of Arbela, rather improbable. One might as well suggest that the Battle of Waterloo was won by the headmaster of Eton.

It could be claimed, with greater justification, that the most influential book on the conduct, as well as on the occurrence of war has been the Holy Bible. It has been testified that the Old Testament

* Michael Howard, *Studies in War and Peace*.
† B. H. Liddell Hart, *Great Captains Unveiled*.
‡ Charles de Gaulle, *The Army of the Future*.

contains useful military information, especially for those who, as circumstances frequently demand, are required to campaign in the area. Lloyd George provided Allenby with a biblical commentary, remarking that he would find it more helpful than any war office manual.

It must remain a matter of opinion whether the wider military influence of the Word of the Lord has been more than a general inspiration. "I can say this of Naseby," said Cromwell after the victory, "that when I saw the enemy drawn up and march in gallant order towards us, and we a company of poor ignorant men, to seek how to order out battle . . . I could (riding alone about my business) but smile out to God in praises, in assurance of victory, because God would, by things that are not, bring to naught the things that are."

Lenin's theoretical military views, it has been officially reiterated, are the foundation of the military science of the Soviet Union. So, too, in China and elsewhere with the writings of Chairman Mao. Yet how relevant are they to the strategical and tactical problems of waging war?

"The more I see of war," Wavell wrote to my father in 1942 when, as Supreme Commander in the Southwest Pacific, he was trying to stem the Japanese advance, "the less I think that general principles of strategy count as compared with administrative problems and the gaining of intelligence. The main principles of strategy, e.g. to attack the other fellow in the flank or rear in preference to the front, to surprise him by any means in one's power and to attack his morale before you attack him physically are really things that every savage schoolboy knows. But it is often outside the power of the general to act as he would have liked owing to lack of adequate resources and I think that military history very seldom brings this out, in fact it is almost impossible that it should do so without a detailed study which is often unavailable. For instance, if Hannibal had another twenty elephants, it might have altered his whole strategy against Italy."

Yet it is undeniable that the intellectual aspect of war, expressed in the principles of strategy and their development, can exert a strong attraction. In one of his earlier books my father referred to the long passage in *The Guermantes Way,* not long published, in which Proust describes and analyzes the attraction of this art. In *The Seven Pillars of Wisdom,* T. E. Lawrence likewise relates his "real" battle to an intellectual strategy—and also provides an interplay with other aesthetic attractions.

The material aspect of war has produced a literature often useful at the time but scarcely memorable. The capabilities of Hannibal's elephants, like the details of the crossbow, are of remote interest. The technical propaganda for—or against—more recent weapon systems becomes as rapidly obsolete as the objects themselves, as anyone who has worked in this field must be aware.

One of the problems which has so far defeated even the detached and informed writer on modern war is, I think, to bring out the essential bearing of complex armament and "logistics" on the conduct of war. Speer's description, published after twenty years in prison as a war criminal, of his time as Minister of Armaments in the greatest war machine then created, brings out the amateurishness with which these aspects were related both to strategical and moral considerations in the supreme direction of total, or what Clausewitz had conceived to be "absolute" war.

"One can only wonder," he says, "at the recklessness and frivolity with which Hitler appointed me to one of those three or four ministries on which the existence of his state depended. Never in my life had I anything to do with military weapons. . . ."*

The moral aspect of war is that which predominantly engages our attention, whether this be the treatment of the laws and ethics of war or of the morale of soldiers and civilians at different levels. War, after all, is not a game. Why do men fight—and sometimes not fight? What was their responsibility? And, in the event, what was to be the determining factor when, as Tolstoy describes in *War and Peace,* "that moment of moral vacillation had come which decides the fate of battles"?

"The soundest strategy," says Lenin, "is to postpone operations until the disintegration of the enemy renders the delivery of the mortal blow both possible and easy."

It is on the moral aspect of war that "intellectual" writers on strategy have, like commanders in retrospect, found it most difficult to avoid being trite. Personal experience of war has, however, often saved such writers from the fatuities in which civilians have indulged.

Such a professional military strategist as Field Marshal Von der Goltz comments that "the possession of a horse furnishes a man in the hour of his greatest danger with the means of saving himself and it cannot be expected of him that he should not avail of it."† Per-

---

* Albert Speer, *Inside the Third Reich.*
† Colmar Von der Goltz, *The Nation in Arms.*

haps he had in mind Frederick the Great's flight at the Battle of Mollwitz.

Most strategists and historians are still inhibited from discussing, in comparative detail and in a particular context of experience, the operation of those moral factors in war whose crucial influence they theoretically acknowledge. Can one seriously question the resolution of one's own commanders and troops, of those of one's allies, or even, perhaps, those of the enemy, whatever may be said about errors of judgment or deficiencies of resource? However, between the Clausewitzian and Tolstoyan poles, there is a moderate or limited view of war, well expressed strategically by Marshal Saxe and more recently satirized by Waugh: "The allies had lately much impeded their advance by the destruction of Monte Cassino, but the price of this sacrilege was being paid by the infantry of the front line. It did not trouble the peace-loving and unambitious officers who were glad to settle in Bari."*

Dealing with lawbreakers who have committed violence, I sometimes wonder at the confidence with which society decides that to kill in some circumstances and not to kill in other circumstances are alike reprehensible—or, in the psychiatric view, abnormal. A number of interesting studies have been made of aggression and war from a psychological point of view. It must be admitted that the lives not only of the "great captains" but of those who have devoted themselves to writing about war offer clinical evidence; a rather high proportion have not only lost their heads, possibly an occupational risk, but have, at some stage, gone *off* their heads.

A psychiatrist, complementing a strategical study of Churchill by my father, concludes that "his inspirational quality owed its dynamic force to the romantic world of phantasy in which he had his true being."†

It may be that the readiness to wage war, at all levels and in all situations, is largely a matter of duty—but this only begs the question. None of the lengthy accounts by his fellow generals, explaining why they would have been more successful if only they had not been let down, is more revealing than that of Field Marshal Keitel, composed —when he knew he was to be hanged—somewhat unprofessionally at

---

* Evelyn Waugh, *Unconditional Surrender.*
† Anthony Storr, *Churchill: Four Faces and the Man.*

Nuremburg and necessarily brief. "The officer's profession is not a liberal profession: a soldier's cardinal virtue is obedience, in other words the very opposite of criticism . . . the so-called 'manic' intellectual does not make a suitable officer while, on the other hand, the one-sided education of the professional soldier described above results in a lack of ability to make a stand against theses which are not part of his real territory."*

Has the pen been any better than the sword in resolving the moral dilemmas of war? "Among the calamities of war," Dr. Johnson wrote, "may be justly numbered the diminution of the love of truth by the falsehoods which interest dictates and credulity encourages. A peace will equally leave the warrior and the relater of wars destitute of employment; and I know not whether more is to be dreaded from streets filled with soldiers accustomed to plunder or from garrets filled with scribblers accustomed to lie."†

The conflict between the Pen and the Sword has been a recurrent theme in the history of war. War leaders and generals and the rank and file have attacked treasonable critics and armchair strategists. Writers, in turn, have attacked stupid and bloodthirsty soldiers. "Men wearing rapiers are afraid of goose-quills," Shakespeare commented; Napoleon observed that "four hostile newspapers were more to be feared than a thousand bayonets."

Yet the antithesis is not clear-cut. Many great war leaders have owed their positions more to the influence of their pens than to any accomplishments on the battlefield. Throughout history, generals have tried to enhance their reputations by the use of their pens or those of their associates and, if Procopius is to be believed, historians have lied in fear of the sword. The sword has frequently been used to destroy the pen and its users; the latter have just as often expedited the work of destruction. There is nothing new in the martial use of the pen—and the printing press.

Trotsky was hunted down and killed with an ice pick while he was writing. In his *History of the Russian Revolution* he had described how he himself had used the pen to destroy. The sword was used by him in an apparently confused and incidental fashion (though the battles are prominently depicted in the "official" version, from which virtually every mention of Trotsky has been wiped out).

---

* *Memoirs of Field Marshal Keitel.*
† Samuel Johnson, *The Idler*, 1758.

"Who can ascertain the truth about a cannon shot fired in the thick of night from a mutinous ship at a Czar's palace where the last government of the possessing classes is going out like an oil-less lamp? But just the same," Trotsky concludes, "the historian will make no mistake if he says that on October 25th not only was the electric current shut off in the government printing plant but an important page was turned in the history of mankind."

The pen has frequently been used as a weapon in war—against one's own warriors. Swift's *Conduct of the Allies* includes an analysis of Britain's historic strategy, but Swift was not writing an academic thesis. He was employed for an immediate purpose in that war.

"While it lasted," Sir Walter Scott comments "it was impossible to dismiss Marlborough without the most awful responsibility, and the only alternative which remained was to render the war unpopular. With this view Swift's *Conduct of the Allies* was published and produced the deepest sensation upon the public mind."* Smollett commented: "That hero who had retrieved the glory of British arms . . . and, as it were, chained victory to his chariot wheels was, in a few weeks, dwindled into an object of contempt and derision."

Moreover, it has been common for great writers on war, as well as many lesser ones, to show marked inconsistency and ambivalence toward their subject. Strategists and historians have changed their opinions in particular instances. About war as a whole, writers have exemplified the contradictions both in their successive works and in their lives. "But what can war but endless war still breed?"† asks Milton, who had indefatigably defended until the last a military regime which put the Irish and its other enemies to the sword.

In *Don Juan* Byron sets out to satirize war, choosing to ridicule Suvorov's campaign against the Turks. "How horrible an example of human nature is this man," wrote Keats, "who has no pleasure left him but to gloat over and jeer at the most awful incidents of life . . . and yet it will fascinate thousands by the very diabolical outrage of their sympathies." Byron, who had already celebrated the struggle of ancient Greece against the Persians, was soon to die, trying to organize, not very successfully, a campaign against these same Turks.

I recall the disruption between my father and some of his literary friends who had been fervent preachers against war, and in some cases conscientious objectors, when, after Russia entered World War

---

* Sir Walter Scott, ed., *Works of Jonathan Swift.*
† John Milton, *To the Lord General Fairfax.*

II they began to write with like fervor—and with utter disregard for the strategical problems—in favor of immediately opening a second front. I also recall the dismay when, during the Blockade of Berlin, in 1948, I was assigned to look after the arrangements for the visit of Bertrand Russell, and it was discovered that the philosopher had chosen this time and place to call for the threat, at least, of a preventive nuclear war.

"The arms race became inevitable unless drastic measures were taken to avoid it," he was to write in his *Autobiography*. "That is why, in late 1948, I suggested that the remedy might be the threat of immediate war by the United States. . . ." He admits that he had hotly denied that he ever made such a suggestion and remarks that it is shameful to deny one's own words.

"To conclude, therefore," Bacon writes to Prince Charles in 1624, "howsoever some schoolmen, otherwise reverend men, yet fitter to guide penknives than swords, seem precisely to stand upon it . . . a just fear will be a just cause of a preventive war; but especially if it be part of the case, that there be a nation that is manifestly detected to aspire to monarchy and new conquests; then other states, assuredly, cannot be justly accused for not staying the first blow, or for not accepting Polyphemus's courtesy, to be the last that shall be eaten up."*

Soldiers have, naturally, sought to diminish the force of war criticism by emphasizing the value of personal experience of war. Frederick the Great made Guichard, a military historian, stand to attention with a full pack for several hours. "You will agree that you can only judge of certain things by comparison. Our friends, the authors, decide things in their study and it is well to correct their ideas by practice. . . . Our Captain will no longer pass judgment, if he ever writes again, so lightly as he did, after the experience he has been put through, which seemed . . . to sadden him somewhat."

"Had Grotius been a commander," Gustavus Adolphus remarked, "he would have seen that his precepts could not be carried out."

Naturally, too, those military writers who have had some war experience tend to support this view. "The Captain of the Hampshire Grenadiers," Gibbon concludes in his *Autobiography*, "has not been useless to the historian of the Roman Empire." Polybius maintained

* Sir Francis Bacon, *On War with Spain.*

that it was impossible to write well on the operations of war if a man had no experience of actual service.

Moreover, the oversimplified antithesis between the Sword and the Pen obscures the truth that some of the most effective pen-pushing has been carried out, not in public exposition of the art of war but in the service of the war "machine." War is fashioned by the materials of written instruction. And in modern war, or preparation for war, most generals themselves are likely to spend much of their careers skillfully pushing a pen in performance of their duties.

Arthur Bryant observes in his biography of Pepys that "more than any other man he evolved the gentle art of bureaucratic defence and offence," and that "when it came to the niceties of battle by administrative correspondence he was their master every time."*

Pepys wrote: "I am loath it should be thought possible that any degree of friendship or other consideration whatever could prevail with me to mislead his Majesty, by one word of mine, to the granting of a thing so extraordinary, so irregular and unjustified by any practice past, and unlikely to be ever imitated in time to come, as this which you have thus contended for of having two of the top flags of England exposed to sea in view of the two greatest rivals of England for Sea Dominion and Glory (I mean the Dutch and French) with no better provision for supporting the honour thereof than six ships, and two of them such as carry not above 190 men and 54 guns between them. And this, too, obtained through mere force of importunity by one who but in September last charged Captain Priestman with turning the King's flag into ridicule in putting up but an unusual swallow-tailed pendant. . . ."

Finally, some of the most notable writing on war has been concerned with the constitutional problems of military power. In modern democracies, as much as in ancient Greece, men are concerned with organizing their defense forces in a way which will not threaten their own society.

Is there a dimension in which the various and contradictory ways in which war has been treated by word and pen, and the ambivalence shown by their use may be reconciled?

Melville had been one of those who, in his earlier works such as *White-Jacket,* had written against war and the abuses which he himself had experienced in a "man-o'-war." In his last short allegorical work, *Billy Budd,* he is "prompted by the sight of the star inserted in

* Sir Arthur Bryant, *Samuel Pepys, Saviour of the Navy.*

the *Victory's* quarterdeck designating the spot where the Great Sailor fell," to muse on the resolution of truth and myth, careful strategy and dramatic art in war.

"If under the presentiment of the most magnificent of all victories, to be crowned by his own glorious death, a sort of priestly motive led him to dress his person in the jewelled vouchers of his own shining deeds; if thus to have adorned himself for the altar and the sacrifice were indeed vainglory, then affectation and fustian is each more heroic line in the great epics and dramas, since in such lines the poet but embodies in verse those exaltations of sentiment that a nature like Nelson's, the opportunity being given, vitalizes into acts."

This anthology, then, is an exploration of various interpretations of war. Inevitably, many selections suffer out of their context. They may, however, be better appreciated when they are set against those of their contemporaries. In this context, too, the obscurer view has its value; Private Wheeler tells us something about Wellington's "grand strategy."

Circumstances have dictated a possibly undue preponderance of writing from the Western and, in particular, from the English-speaking world. This project was surely not conceived under the impression that we had more than our national share of military or literary genius—though it may be that the English have got themselves involved in an undue number of wars. One may speculate, too, on the nature of a political or social climate favorable to the literary expression of an interest in war—in poetry or polemic.

Numerous accounts of World War II, from those who played some part in it, are still emerging. There has been a massive bombardment—and many resounding claims. "I hear that my generals are selling their Lives dearly," Churchill is said to have commented in old age. He was hardly in a position to complain. "In fifty years time your name will be a household word," Field Marshal Viscount Allenby of Megiddo remarked to Colonel T. E. Lawrence after World War I. "To find out about Allenby, they will have to go to the War Museum." He was not far wrong.

Thucydides, himself an unsuccessful admiral, commented that "many badly conceived enterprises have had the luck to be successful because the enemy has shown an even smaller degree of intelligence." However, a few hundred years later Josephus, an ex-general with no mean powers of self-congratulation, explained in the introduction to his *History of the Jewish War,* the importance of giving the enemy his due. "Yet the writers I have in mind claim to be writing history, though besides getting all their facts wrong they seem to miss the target altogether. For they wish to establish the greatness of the Romans while all the time disparaging and deriding the actions of the Jews. But I do not see how men can prove themselves great by overcoming feeble opponents." Such advice may have fostered a mutual admiration society.

If, when it comes to World War II, it is difficult to maintain a sense of historical proportion, it may also be invidious—and premature—to consider the influence on war of postwar writing on the nuclear deterrent and other matters. As Plato said, "only the dead have seen the end of the war." And only the dead, perhaps, should be wisely included in an anthology of war.

Editorial Note

In reproducing writing from many sources I have kept closely to the text, even where it would seem that the author has slipped in language or in reference. With translations, especially from the ancient classics, and with regard to technical terms, there may be arguments about the correct rendering. In general I have used the standard translations, where available; these have influenced people in the past, though modern scholarship may, in some cases, have improved on them. With old English I have modernized the spelling, though not the style or, I trust, the sense.

I.
THE
ANCIENT
WORLD

War is the father of all things.
—HERACLITUS

Book of Judges

So the people took victuals in their hand, and their trumpets: and he sent all the rest of Israel every man unto his tent, and retained those three hundred men: and the host of Midian was beneath him in the valley.

And it came to pass the same night, that the Lord said unto him, Arise, get thee down unto the host; for I have delivered it into thine hand. . . .

So Gideon, and the three hundred men that were with him, came unto the outside of the camp in the beginning of the middle watch; and they had but newly set the watch; and they blew the trumpets, and brake the pitchers that were in their hands. And the three companies blew the trumpets, and brake the pitchers, and held the lamps in their left hands, and the trumpets in their right hands to blow withal; and they cried, The sword of the Lord, and of Gideon.

And they stood every man in his place round about the camp: and all the host ran, and cried, and fled.

And the three hundred blew the trumpets, and the Lord set every man's sword against his fellow, even throughout all his host: and the host fled to Bethshittah in Zererath, and to the border of Abelmeholah, unto Tabbath.

THUCYDIDES:

FROM THE

History of the Peloponnesian War

"It will be enough for me, however, if these words of mine are judged useful by those who want to understand clearly the events which happened in the past and which (human nature being what it is) will, at some time or other and in much the same ways, be repeated in the future. My work is not a piece of writing designed to meet the taste of an immediate public, but was done to last forever."

Thucydides was born in Athens in 460 B.C. and was elected a strategus in 424. As such he commanded an Athenian fleet during the Peloponnesian War against the Spartans. He was recalled and exiled for his responsibility in the loss of Amphipolis after a sudden winter attack. During twenty years of exile he wrote his history, returning to Athens after its fall in 404. It is believed that he died by violence in 399.

The Athenians then held an assembly in order to debate the matter, and decided to look into the whole question once and for all and then to give Sparta her answer. Many speakers came forward and opinions were expressed on both sides, some maintaining that war was necessary and others saying that the Megarian decree should be revoked and should not be allowed to stand in the way of peace. Among the speakers was Pericles, the son of Xanthippus, the leading man of his time among the Athenians and the most powerful both in action and in debate. His advice was as follows: "Athenians," he said, "my views are the same as ever: I am against making any concessions to the Peloponnesians, even though I am aware that the enthusiastic state of mind in which people are persuaded to enter upon a war is not retained when it comes to action, and that people's minds are altered by the course of events. Nevertheless I see that on this occasion I must give you exactly the same advice as I have given in the past, and I call upon those of you who are persuaded by my words to give your full support to these resolutions which we are making all together, and to abide by them even if in some respect or other we find ourselves in difficulty; for, unless you do so, you will be able to claim no credit for intelligence when things go well with us. There is often no more logic in the course of events than there is in the plans of men, and this is why we usually blame our luck when things happen in ways that we did not expect.

"It was evident before that Sparta was plotting against us, and now it is even more evident. It is laid down in the treaty that differences between us should be settled by arbitration, and that, pending arbitration, each side should keep what it has. The Spartans have never once asked for arbitration, nor have they accepted our offers to submit to it. They prefer to settle their complaints by war rather than by peaceful negotiations, and now they come here not even making protests, but trying to give us orders. They tell us to abandon the siege of Potidaea, to give Aegina her independence, and to revoke the Megarian decree. And finally they come to us with a proclamation that we must give the Hellenes their freedom.

"Let none of you think that we should be going to war for a trifle if we refuse to revoke the Megarian decree. It is a point they make much of, and say that war need not take place if we revoke this decree; but, if we do go to war, let there be no kind of suspicion in your hearts that the war was over a small matter. For you this trifle is both the assurance and the proof of your determination.

"If you give in, you will immediately be confronted with some greater demand, since they will think that you only give way on this point through fear. But if you take a firm stand you will make it clear to them that they have to treat you properly as equals. And now you must make up your minds what you are going to do—either to give way to them before being hurt by them, or, if we go to war—as I think we should do—to be determined that, whether the reason put forward is big or small, we are not in any case going to climb down nor hold our possessions under a constant threat of interference. When one's equals, before resorting to arbitration, make claims on their neighbors and put those claims in the form of commands, it would still be slavish to give in to them, however big or however small such claims may be.

"Now, as to the war and to the resources available to each side, I should like you to listen to a detailed account and to realize that we are not the weaker party. The Peloponnesians cultivate their own land themselves; they have no financial resources either as individuals or as states; then they have no experience of fighting overseas, nor of any fighting that lasts a long time, since the wars they fight against each other are, because of their poverty, short affairs. Such people are incapable of often manning a fleet or often sending out an army, when that means absence from their own land, expense from their own funds and, apart from this, when we have control of the sea. And wars are paid for by the possession of reserves rather than by a sudden increase in taxation. Those who farm their own land, moreover, are in warfare more anxious about their money than their lives; they have a shrewd idea that they themselves will come out safe and sound, but they are not at all sure that all their money will not have been spent before then, especially if, as is likely to happen, the war lasts longer than they expect.

"In a single battle the Peloponnesians and their allies could stand up to all the rest of Hellas, but they cannot fight a war against a power unlike themselves, so long as they have no central deliberative authority to produce quick decisive action, when they all have equal votes, though they all come from different nationalities and every one of these is mainly concerned with its own interests—the usual result of which is that nothing gets done at all, some being particularly anxious to avenge themselves on an enemy and others no less anxious to avoid coming to any harm themselves. Only after long intervals do they meet together at all, and then they devote only a fraction of their time to their general interests, spending most of it on arranging their own separate affairs. It never occurs to any of them that the apathy of one will damage the interests of all. Instead each state thinks that the responsibility for its future belongs to someone else, and so, while everyone has the same idea privately, no one notices that from a general point of view things are going downhill.

"But this is the main point: they will be handicapped by lack of money and delayed by the time they will have to take in procuring it. But in war opportunity waits for no man.

"Then we have nothing to fear from their navy, nor need we be alarmed at the prospect of their building fortifications in Attica. So far as that goes, even in peace time it is not easy to build one city strong enough to be a check upon another; and this would be a much harder thing to accomplish in enemy territory and faced with our own fortifications, which are just as strong as anything that they could build. While if they merely establish some minor outpost, they could certainly do some harm to part of our land by raiding and by receiving deserters, but this could by no means prevent us from retaliating by the use of our sea-power and from sailing to their territory and building fortifications there. For we have acquired more experience of land fighting through our naval operations than they have of sea fighting through their operations on land. And as for seamanship, they will find that a difficult lesson to learn. You yourselves have been studying it ever since the end of the Persian wars, and have still not entirely mastered the subject. How, then, can it be supposed that they could ever make much progress? They are farmers, not sailors, and in addition to that they will never get a chance of practising, because we shall be blockading them with strong naval forces. Against a weak blockading force they might be prepared to take a risk, bolstering up their ignorance by the thought of their numbers, but if they are faced with a large fleet they will not venture out, and so lack of practice will make them even less skilful than they were, and lack of skill will make them even less venturesome. Seamanship, just like anything else, is an art. It is not something that can be picked up and studied in one's spare time; indeed, it allows one no spare time for anything else.

"Sea-power is of enormous importance. Look at it this way. Suppose we were an island, would we not be absolutely secure from attack? As it is we must try to think of ourselves as islanders; we must abandon our land and our houses, and safeguard the sea and the city. We must not, through anger at losing land and homes, join battle with the greatly superior forces of the Peloponnesians. If we won a victory, we should still have to fight them again in just the same numbers, and if we suffered a defeat, we should at the same time lose our allies, on whom our strength depends, since they will immediately revolt if we are left with insufficient troops to send against them. What we should lament is not the loss of houses or of land, but the loss of men's lives. Men come first; the rest is the fruit of their labour. And if I thought I could persuade you to do it, I would urge you to go out and lay waste your property with your own hands and show the Peloponnesians that it is not for the sake of this that you are likely to give in to them.

"I could give you many other reasons why you should feel confident in ultimate victory, if only you will make up your minds not to add to the empire while the war is in progress, and not to go out of your way to involve yourselves in new perils. What I fear is not the enemy's strategy, but our own mistakes. However, I shall deal with all this on another occasion when words and action will go together. For the present I recommend that we send back the Spartan ambassadors with the following answer: that we will give Megara access to our market and our ports, if at the same time Sparta exempts us and our allies from the operation of her orders for the expulsion of aliens (for in the treaty there is no clause forbidding either those orders of hers or our decree against Megara); that we will give their independence to our allies if they had it at the time that we made the treaty and when the Spartans also allow their own allies to be independent and to have the kind of government each wants to have rather than the kind of government that suits Spartan interests. Let us say, too, that we are willing, according to the terms of the treaty, to submit to arbitration, that we shall not start the war, but that we shall resist those who do start it. This is the right reply to make and it is the reply that this city of ours ought to make. We must realize that this war is being forced upon us, and the more readily we accept the challenge the less eager to attack us will our opponents be. We must realize, too, that, both for cities and for individuals, it is from the greatest dangers that the greatest glory is to be won. When our fathers stood against the Persians they had no such resources as we have now; indeed, they abandoned even what they had, and then it was by wisdom rather than by good fortune, by daring rather than by material power, that they drove back the foreign invasion and made our city what it is today. We must live up to the standard they set: we must resist our enemies in any and every way, and try to leave to those who come after us an Athens that is as great as ever."

—REX WARNER
(translator)

XENOPHON:

FROM THE

Anabasis

Xenophon was born in Athens in 430 B.C. and probably served during the later campaigns of the Peloponnesian War. He became a disciple of Socrates and is one of the main sources of information about the philosopher. He went to Asia Minor as a mercenary soldier, joining Cyrus the Younger in his unsuccessful struggle for the Persian throne. After a long retreat, during which he took command, and about which he wrote his Anabasis, he joined the Thracian and then the Spartan forces. He managed to make his fortune as a soldier, settled at Corinth for a time, and may have returned to Athens, from which he had been exiled, before his death about 350.

During the retreat—or escape—of the Ten Thousand from Persia to the Mediterranean coast, the Greek mercenaries under Xenophon capture a mountain pass.

Next came a two-days' march of thirty miles. At the pass which led down into the plain there were Chalybes, Taochi and Phasians to bar their way, and, when Chirisophus saw that the enemy was holding the pass, he came to a halt, keeping about three miles away from them, so as not to approach them while marching in column. He sent orders to the other officers to bring up their companies on his flank, so that the army should be in line. When the rearguard had got into position he called a meeting of the generals and captains, and spoke as follows: "As you see, the enemy are holding the pass over the mountain. Now is the time to decide what is the best method of dealing with them. What I suggest is that we give orders to the troops to have a meal, and meanwhile decide whether it is best to cross the mountain today or tomorrow."

"I think, on the other hand," said Cleanor, "that we should get ready for battle and make an attack, as soon as we have finished our meal. My reason is that, if we let this day go by, the enemy who are now watching us will gain confidence and if they do, others will probably join them in greater numbers."

Xenophon spoke next, and said: "This is my view. If we have to fight a battle, what we must see to is how we may fight with the greatest efficiency. But if we want to get across the mountain with the minimum of inconvenience, then, I think, what we must consider is how to ensure that our casualties in dead and wounded are as light as possible. The mountain, so far as we can see, extends for more than six miles, but except just for the part on our road, there is no evidence anywhere of men on guard against us. It would be a much better plan, then, for us to try to steal a bit of the undefended mountain from them when they are not looking, and to capture it from them, if we can, by taking the initiative, than to fight an action against a strong position and against troops who are waiting ready for us.

"It is much easier to march uphill without fighting than to march on the level when one has enemies on all sides; and one can see what is in front of one's feet better by night, when one is not fighting, than by day, if one is; and rough ground is easier for the feet, if one is not fighting as one marches, than level ground is, when there are weapons flying round one's head. I do not think that it is impossible for us to steal this ground from them. We can go by night, so as to be out of their observation; and we can keep far enough away from them to give them no chance of hearing us. And I would suggest that, if we make a feint at attacking here, we should find the rest of the mountain even less defended, as the enemy would be likely to stay here in a greater concentration.

"But I am not the person who ought to be talking about stealing. I gather that you Spartans, Chirisophus—I mean those of you who belong to the Peers—study how to steal from your earliest boyhood, and think that so far from it being a disgrace it is an actual distinction to steal anything that is not forbidden by law. And, so that you may become expert thieves and try to get away with what you steal, it is laid down by law that you get a beating if you are caught stealing. Here then is an excellent opportunity for you to give an exhibition of the way in which you were brought up, and to preserve us from blows, by seeing to it that we are not caught stealing our bit of mountain."

"Well," said Chirisophus, "what I have gathered about you Athenians is that you are remarkably good at stealing public funds, even though it is a very risky business for whoever does so; and your best men are the greatest experts at it, that is if it is your best men who are considered the right people to be in the government. So here is a chance for you too to give an exhibition of the way in which you were brought up."

"Then," said Xenophon, "I am prepared, as soon as we have had our meal, to take the rearguard and go to seize the position in the mountains. I have got guides already, as my light troops ambushed and made prisoners of a few of the natives who have been following behind to pick up what they could. I have also been informed by them that the mountains are not impassable: they provide pasture for goats and cattle. If, therefore, we once get hold of a part of the range, there will be a possible route for our baggage animals as well. I do not expect either that the enemy will stand their ground when they see that we are holding the heights and on a level with them, as they show no willingness at the moment to come down on to a level with us."

"But why," said Chirisophus, "should you go and leave vacant the command of the rearguard? It would be better to send others, that is if some good soldiers do not come forward as volunteers."

Then Aristonymus of Methydria, a commander of hoplites, and Aristeas of Chios, and Nicomachus of Oeta, commanders of light infantry, came forward, and it was agreed that they would light a number of fires as soon as they had seized the heights. When this was settled they had their meal, and afterwards Chirisophus led the army forward about a mile in the direction of the enemy, so as to give the impression that it was at this point that he intended to attack.

When they had had supper and it became dark, the troops detailed for the job set off and seized the mountain height, while the others rested where they were. As soon as the enemy realized that the heights had been occupied, they were on the look-out and kept a number of fires burning through the night. At daybreak Chirisophus offered sacrifices and then advanced on the road, while the troops who had seized the mountain ridge made an attack along the heights. Most of the enemy stood their ground at the pass, but part of them went to engage the troops on the heights. However, before the main bodies came to close quarters, the troops on the heights were in action and the Greeks were winning and driving the enemy back.

At the same moment in the plain the Greek peltasts advanced at the double against the enemy's battle line, and Chirisophus with the hoplites followed at a quick march behind. However, when the enemy guarding the road saw that their troops higher up were being defeated, they took to flight. Not many of them were killed, but a very great number of shields were captured. The Greeks cut these shields up with their swords, and so made them useless. When they reached the summit, they offered sacrifices and set up a trophy.

—S. A. HANDFORD
(translator)

ARISTOTLE:

FROM

Politics

*Aristotle was born in Greece in 384 B.C. and at seventeen
joined Plato's Academy in Athens. In 342 he became tutor
to the King of Macedon's son, the future Alexander the
Great. After seven years, when Alexander set out on his
conquests, Aristotle returned to Athens and devoted himself increasingly to scientific studies. After a revolt he fled to
Chalcis and died there in 322.*

Just as there are four chief divisions of the mass of the population—farmers, mechanics, shopkeepers and day-laborers—so there are
also four kinds of military force—cavalry, heavy infantry, light-armed
troops and the navy. Where a territory is suitable for the use of cavalry, there is a favorable ground for the construction of a strong form
of oligarchy: the inhabitants of such a territory need a cavalry force
for security, and it is only men of large means who can afford to
breed and keep horses. Where a territory is suitable for the use of
heavy infantry, the next and less exclusive variety of oligarchy is natural; service in the heavy infantry is a matter for the well-to-do rather
than for the poor. Light-armed troops and the navy are drawn from
the mass of the people and are thus wholly on the side of democracy
—with the light-armed troops and naval forces as large as they are,
the oligarchical side is generally worsted in any civil dispute.

This situation should be met, and remedied, by following the
practice of some military commanders who combine an appropriate
number of light-armed troops with the cavalry and infantry. The
reason why the masses can defeat the wealthier classes, in any civil

dissension, is that a light-armed and mobile force finds it easy to cope with a force of cavalry and heavy infantry. An oligarchy which builds up a force of light-armed men exclusively from the masses is thus only building up a challenge to itself.

SUN TZU:

FROM

The Art of War

Sun Tzu has often been described as the earliest classic writer on war and for centuries his writings have been studied in China and Japan. Recent research indicates that his work on The Art of War *was written somewhat later than supposed, about 400–320 B.C. Sun Tzu was a general in the service of the King of Wu and is said to have captured Ying, the capital of the Chu'u state, and to have defeated the Ch'i and Chin states to the north.*

Generally in war the best policy is to take a state intact; to ruin it is inferior to this.

To capture the enemy's is better than to destroy it; to take intact a battalion, a company or a five-man squad is better than to destroy them. For to win one hundred victories in one hundred battles is not the acme of skill. To subdue the enemy without fighting is the acme of skill.

Thus, what is of supreme importance in war is to attack the enemy's strategy.

Next best is to disrupt his alliances.

The next best is to attack his army.

The worst policy is to attack cities. Attack cities only when there is no alternative.

To prepare the shielded wagons and make ready the necessary arms and equipment requires at least three months; to pile up earthen ramps against the walls an additional three months are needed.

If the general is unable to control his impatience and orders his troops to swarm up the walls like ants, one third of them will be killed without taking the city. Such is the calamity of these attacks.

Thus, those skilled in war subdue the enemy's army without battle. They capture his cities without assaulting them and overthrow his state without protracted operations.

Your aim must be to take everything intact. Thus your troops are not worn out and your gains will be complete. This is the art of offensive strategy.

Consequently, the art of using troops is this: When ten to the enemy's one surround him; When five times his strength, attack him; If double his strength, divide him; If equally matched, you may engage him; If weaker numerically, be capable of withdrawing; And if in all respects unequal, be capable of eluding him, for a small force is but booty for one more powerful.

Now the general is the protector of the state. If this protection is all-embracing, the state will surely be strong; if defective, the state will certainly be weak.

Now there are three ways in which a ruler can bring misfortune upon his army:

When ignorant that the army should not advance, to order an advance, or ignorant that it should not retire, to order a retirement. This is described as "hobbling the army."

When ignorant of military affairs, to participate in their administration.

This causes the officers to be perplexed.

When ignorant of command problems to share in the exercise of responsibilities. This engenders doubts in the minds of officers. If the army is confused and suspicious, neighboring rulers will cause trouble. This is what is meant by the saying: "A confused army leads to another's victory."

Now there are five circumstances in which victory may be predicted: He who knows when he can fight and when he cannot will be victorious.

He who understands how to use both large and small forces will be victorious.

He whose ranks are united in purpose will be victorious.

He who is prudent and lies in wait for an enemy who is not, will be victorious.

He whose generals are able and not interfered with by the sovereign will be victorious.

It is in these five matters that the way to victory is known.

Therefore I say: "Know the enemy and know yourself"; in a hundred battles you will never be in peril.

When you are ignorant of the enemy but know yourself, your chances of winning or losing are equal.

If ignorant both of the enemy and of yourself, you are certain in every battle to be in peril.

* * *

Anciently the skilful warriors first made themselves invincible and awaited the enemy's moment of vulnerability.

Invincibility depends on one's self; the enemy's vulnerability on him.

It follows that those skilled in war can make themselves invincible but cannot cause an enemy to be certainly vulnerable. Therefore it is said that one may know how to win, but cannot necessarily do so.

Invincibility lies in the defense; the possibility of victory in the attack.

One defends when his strength is inadequate; he attacks when it is abundant.

The experts in defense conceal themselves as under the ninefold earth; those skilled in attack move as from above the ninefold heavens. Thus they are capable both of protecting themselves and of gaining a complete victory.

To forsee a victory which the ordinary man can forsee is not the acme of skill.

To triumph in battle and be universally acclaimed "expert" is not the acme of skill, for to lift an autumn down requires no great strength; to distinguish between the sun and moon is no test of vision; to hear the thunderclap is no indication of acute hearing. Anciently those called skilled in war conquered an enemy easily conquered.

And therefore the victories won by a master of war gain him neither reputation for wisdom nor merit for valour.

For he wins his victories without erring. "Without erring" means that whatever he does insures his victory; he conquers an enemy already defeated.

Therefore the skilful commander takes up a position in which he cannot be defeated and misses no opportunity to master his enemy. Thus a victorious army wins its victories before seeking battle; an army destined to defeat fights in the hope of winning. It is because of disposition that a victorious general is able to make his people fight with the effect of pent-up waters which, suddenly released, plunge into a bottomless abyss.

* * *

Generally, management of many is the same as management of few. It is a matter of organization.

And to control many is the same as to control few. This is a matter of formation and signals.

That the army is certain to sustain the enemy's attack without suffering defeat is due to the operations of the extraordinary and the normal forces.

Troops thrown against the enemy as a grindstone against eggs is an example of a solid acting upon a void.

Generally, in battle, use the normal force to engage; use the extraordinary to win.

Now the resources of those skilled in the use of extraordinary forces are as infinite as the heavens and earth; as inexhaustible as the flow of the great rivers.

For they end and recommence; cyclical, as are the movements of the sun and moon. They die away and are reborn; recurrent, as are the passing seasons.

The musical notes are only five in number but their melodies are so numerous that one cannot hear them all.

The primary colours are only five in number but their combinations are so infinite that one cannot visualize them all.

The flavours are only five in number but their blends are so various that one cannot taste them all.

In battle there are only the normal and extraordinary forces, but their combinations are limitless; none can comprehend them all. For these two forces are mutually reproductive; their interaction as endless as that of interlocked rings. Who can determine where one ends and the other begins?

When torrential water tosses boulders, it is because of its momentum; When the strike of a hawk breaks the body of its prey, it is because of timing.

Thus the momentum of one skilled in war is overwhelming and his attack precisely regulated.

His potential is that of the fully drawn crossbow; his timing, the release of the trigger.

In the tumult and uproar the battle seems chaotic, but there is no disorder; the troops appear to be milling about in circles but cannot be defeated.

Apparent confusion is a product of good order; apparent cowardice, of courage; apparent weakness, of strength.

Order or disorder depends on organization; courage or cowardice on circumstance; strength or weakness on dispositions.

Thus, those skilled at making the enemy move do so by creating a situation to which he must conform; they entice him with something he is certain to take, and with lures of ostensible profit they await him in strength.

Therefore a skilled commander seeks victory from the situation and does not demand it of his subordinates.

He selects his men and they exploit the situation.

—GENERAL S. GRIFFITHS
(translator)

POLYBIUS:

FROM

Histories

Polybius was born in Achaea in 200 B.C. and was deported to Rome as a youth. There he became tutor to Publius Scipio Aemilianus and accompanied him to Carthage in the Third Punic War. He was present when the city fell. In writing his History *he visited the battlefields and had access to the Scipio archives. He is also said to have made a voyage into the Atlantic. He returned to Greece to negotiate a settlement between Achaea and Rome after renewed disturbances and died there in 118.*

Hannibal's great victory over the Romans at Cannae (216 B.C.).

When he took over the command on the following day, as soon as the sun was above the horizon, Gaius Terentius got the army in motion from both the camps. Those from the larger camp he drew up in order of battle, as soon as he had got them across the river, and bringing up those of the smaller camp he placed them all in the same line, selecting the south as the aspect of the whole. The Roman horse he stationed on the right wing along the river, and their foot next them in the same line, placing the maniples, however, closer together than usual, and making the depth of each maniple several times greater than its front. The cavalry of the allies he stationed on the left wing, and the light-armed troops he placed slightly in advance of the whole army, which amounted with its allies to eighty thousand infantry and a little more than six thousand horse.

At the same time Hannibal brought his Balearic slingers and

spearmen across the river, and stationed them in advance of his main body; which he led out of their camp, and, getting them across the river at two spots, drew them up opposite the enemy. On his left wing, close to the river, he stationed the Iberian and Celtic horse opposite the Roman cavalry; and next to them half the Libyan heavy-armed foot; and next to them the Iberian and Celtic foot; next, the other half of the Libyans, and, on the right wing, the Numidian horse. Having now got them all into line he advanced with the central companies of the Iberians and Celts; and so arranged the other companies next these in regular gradations, that the whole line became crescent-shaped, diminishing in depth towards its extremities: his object being to have his Libyans as a reserve in the battle, and to commence the action with his Iberians and Celts.

The armour of the Libyans was Roman, for Hannibal had armed them with a selection of the spoils taken in previous battles. The shield of the Iberians and Celts was about the same size, but their swords were quite different. For that of the Roman can thrust with as deadly effects as it can cut, while the Gallic sword can only cut, and that requires some room. And the companies coming alternately—the naked Celts, and the Iberians with their short linen tunics bordered with purple stripes, the whole appearance of the line was strange and terrifying. The whole strength of the Carthaginian cavalry was ten thousand, but that of their foot was not more than forty thousand, including the Celts. Aemilius commanded on the Roman right, Gaius Terentius on the left, Marcus Atilius and Gnaeus Servilius, the Consuls of the previous year, on the centre. The left of the Carthaginians was commanded by Hasdrubal, the right by Hanno, the centre by Hannibal in person, attended by his brother Mago. And as the Roman line faced the south, as I said before, and the Carthaginian the north, the rays of the rising sun did not inconvenience either of them.

The battle was begun by an engagement between the advanced guard of the two armies; and at first the affair between these light-armed troops was indecisive. But as soon as the Iberian and Celtic cavalry got at the Romans, the battle began in earnest, and in the true barbaric fashion: for there was none of the usual formal advance and retreat; but when they once got to close quarters, they grappled man to man, and, dismounting from their horses, fought on foot. But when the Carthaginians had got the upper hand in this encounter and killed most of their opponents on the ground—because the Romans all maintained the fight with spirit and determination—and began chasing the remainder along the river, slaying as they went and giving no quarter; then the legionnaries took the place of the light-armed and closed with the enemy. For a short time the Iberian and Celtic lines stood their ground and fought gallantly; but, presently overpowered by the weight of the heavy-armed lines, they gave way and retired to the rear, thus breaking up the crescent. The Roman maniples followed with spirit, and easily cut their way through the enemy's line; since the Celts had been drawn up in a thin line, while the Romans had closed up from the wings towards the centre and the point of danger. For the two wings did not come into action at the same time as the centre: but the centre was first engaged, because the Gauls, having been stationed on the arc of the crescent, had come into contact with the enemy long before the wings, the convex of the crescent being towards the enemy. The Romans, however, going in pursuit of these troops, and hastily closing in towards the centre and the part of the enemy which was giving ground, advanced so far, that the Libyan heavy-armed troops on either wing got on their flanks. Those on the right, facing to the left, charged from the right upon the Roman flank; while those who were on the left wing faced to the right, and, dressing by the left, charged their right flank, the exigency of the moment suggesting to them what they ought to do. Thus it came about, as Hannibal had planned, that the Romans were caught between two hostile lines of Libyans—thanks to their impetuous pursuit of the Celts. Still they fought, though no longer in line, yet singly, or in maniples, which faced about to meet those who charged them on the flanks.

Though he had been from the first on the right wing, and had taken part in the cavalry engagement, Lucius Aemilius still survived. Determined to act up to his own exhortatory speech, and seeing that the decision of the battle rested mainly on the legionnaries, riding up to the centre of the line he led the charge himself, and personally grappled with the enemy, at the same time cheering on and exhorting his soldiers to the charge. Hannibal, on the other side, did the same, for he too had taken his place on the centre from the commencement. The Numidian horse on the Carthaginian right were meanwhile charging the cavalry on the Roman left; and though, from the peculiar nature of their mode of fighting, they neither inflicted nor received much harm, they yet rendered the enemy's horse useless by keeping them occupied, and charging them first on one side and then on another. But when Hasdrubal, after all but annihilating the cavalry by the river, came from the left to the support of the Numidians, the Roman allied cavalry, seeing his charge approaching, broke and fled.

At this point Hasdrubal appears to have acted with great skill and discretion. Seeing the Numidians to be strong in numbers, and more effective and formidable to troops that had once been forced from their ground, he left the pursuit to them; while he himself hastened to the part of the field where the infantry were engaged, and brought his men up to support the Libyans. Then, by charging the Roman legions on the rear, and harassing them by hurling squadron after squadron upon them at many points at once, he raised the spirits of the Libyans, and dismayed and depressed those of the Romans. It was at this point that Lucius Aemilius fell, in the thick of the fight, covered with wounds: a man who did his duty to his country at that last hour of his life, as he had throughout its previous years, if any man ever did. As long as the Romans could keep an unbroken front, to turn first in one direction and then in another to meet the assaults of the enemy, they held out; but the outer files of the circle continually falling, and the circle becoming more and more contracted, they at last were all killed on the field. . . .

Such was the end of the battle of Cannae, in which both sides fought with the most conspicuous gallantry, the conquered no less than the conquerors. This is proved by the fact that, out of six thousand horse, only seventy escaped with Gaius Terentius to Venusia, and about three hundred of the allied cavalry to various towns in the neighbourhood. Of the infantry ten thousand were taken prisoners in fair fight, but were not actually engaged in the battle: of those who were actually engaged only about three thousand perhaps escaped to the towns of the surrounding district; all the rest died nobly, to the number of seventy thousand, the Carthaginians being on this occasion, as on previous ones, mainly indebted for their victory to their superiority in cavalry: a lesson to posterity that in actual war it is better to have half the number of infantry, and the superiority in cavalry, than to engage your enemy with an equality in both. On the side of Hannibal there fell four thousand Celts, fifteen hundred Iberians and Libyans, and about two hundred horse.

JULIUS CAESAR:

FROM

Commentaries

*Julius Caesar was born in Rome in 102 B.C. After holding a
succession of military and political posts, he was appointed
governor of Trans-Alpine Gaul and embarked on the subjugation of the country up to the Rhine. In 55 B.C. he made
his first landing in Britain and, during the following years,
crushed the revolt by Vercingetorix and other chiefs in
Gaul. The civil war which broke out in 49 B.C. led to the
final defeat and death of Pompey and Caesar's mastery of
the Roman world. He was assassinated by a group of con-
spirators in 44 B.C.*

*The second invasion of Britain (54 B.C.) during the Gallic
Wars.*

Caesar took with him five legions and the remaining two thousand cavalry, and putting out about sunset was at first carried on his
way by a light southwesterly breeze. But about midnight the wind
dropped, with the result that he was driven far out of his course by
the tidal current and at daybreak saw Britain left behind on the port
side. When the set of the current changed he went with it, and rowed
hard to make the part of the island where he had found the best land-
ing-places the year before. The soldiers worked splendidly, and by
continuous rowing enabled the heavily laden transports to keep up
with the warships. When the whole fleet reached Britain about midday, no enemy was to be seen. Caesar discovered afterwards from
prisoners that, although large numbers had assembled at the spot,
they were frightened by the sight of so many ships and had quitted
the shore to conceal themselves on higher ground.

Caesar disembarked his army and chose a suitable spot for a camp. On learning from prisoners where the enemy were posted, he left ten cohorts and three hundred cavalry on the coast to guard the fleet and marched against the Britons shortly after midnight, feeling little anxiety about the ships because he was leaving them anchored on an open shore of soft sand. The fleet and its guard were put under the command of Quintus Atrius. A night march of about twelve miles brought Caesar in sight of the enemy, who advanced to a river with their cavalry and chariots, and tried to bar his way by attacking from a position on higher ground.

Repulsed by his cavalry they hid in the woods, where they occupied a well-fortified post of great natural strength, previously prepared, no doubt, for some war among themselves, since all the entrances were blocked by felled trees laid close together. Scattered parties made skirmishing attacks out of the woods, trying to prevent the Romans from penetrating their defences. But the soldiers of the 7th legion, locking their shields together over their heads and piling up earth against the fortifications, captured the place and drove them out of the woods at the cost of only a few men wounded. Caesar forbad them to pursue far, however, because he did not know the ground, and because he wanted to devote the few remaining hours of the day to the fortification of his camp.

The next morning he sent out a force of infantry and cavalry in three columns to pursue the fleeing enemy. They had advanced some way and were in sight of the nearest fugitives, when dispatch-riders brought news from Atrius of a great storm in the night, by which nearly all the ships had been damaged or cast ashore; the anchors and cables had not held, and the sailors and their captains could not cope with such a violent gale, so that many vessels were disabled by running foul of one another.

Caesar at once ordered the legions and cavalry to be halted and recalled. He himself went back to the beach, where with his own eyes he saw pretty much what the messengers and the dispatch described. About forty ships were a total loss; the rest looked as if they could be repaired at the cost of much trouble. Accordingly he called out all the skilled workmen from the legions, sent to the continent for more, and wrote to tell Labienus to build as many ships as possible with the troops under his command. Further, although it was a task involving enormous labour, he decided that it would be best to have all the ships beached and enclosed together with the camp by one fortification. This work, although it was continued day and night, took some ten days to complete.

As soon as the ships were hauled up and the camp strongly fortified, Caesar left the same units as before to guard them, and returned to the place from which he had come. On arriving there he found that larger British forces had now been assembled from all sides by Cassivellaunus, to whom the chief command and direction of the campaign had been entrusted by common consent. Cassivellaunus' territory is separated from the maritime tribes by a river called the Thames, and lies about seventy-five miles from the sea. Previously he had been continually at war with the other tribes, but the arrival of our army frightened them into appointing him their supreme commander.

The British cavalry and charioteers had a fierce encounter with our cavalry on the march, but our men had the best of it everywhere and drove them into the woods and hills, killing a good many, but also incurring some casualties themselves by a too eager pursuit. The enemy waited for a time, then, while our soldiers were off their guard and busy fortifying the camp, suddenly dashed out of the woods, swooped upon the outpost on duty in front of the camp, and started a violent battle. Caesar sent two cohorts—the first of their respective legions—to the rescue, and these took up a position close together; but the men were unnerved by the unfamiliar tactics, and the enemy very daringly broke through between them and got away unhurt. That day Quintus Laberius Durus, a military tribune, was killed. The attack was eventually repulsed by throwing in some more cohorts.

Throughout this peculiar combat, which was fought in front of the camp in full view of everyone, it was seen that our troops were too heavily weighted by their armour to deal with such an enemy: they could not pursue them when they retreated, and dared not get separated from their standards. The cavalry, too, found it very dangerous work fighting the charioteers; for the Britons would generally give ground on purpose, and after drawing them some distance from the legions would jump down from their chariots and fight on foot, with the odds in their favour. In engaging their cavalry our men were not much better off: their tactics were such that the danger was exactly the same for both pursuers and pursued. A further difficulty was that they never fought in close order, but in very open formation, and had reserves posted here and there; in this way the various groups covered one another's retreat, and fresh troops replaced those who were tired.

Next day the enemy took up a position on the hills at a distance from the camp. They showed themselves now only in small parties and harassed our cavalry with less vigour than the day before. But at midday, when Caesar had sent three legions and all the cavalry on a foraging expedition under his general Gaius Trebonius, they suddenly swooped down on them from all sides, pressing their attack right up to the standards of the legions. The legionnaries drove them off by a strong counter-attack, and continued to pursue until the cavalry, emboldened by the support of the legions which they saw close behind them, made a charge that sent the natives flying headlong. A great many were killed, and the rest were given no chance of rallying or making a stand or jumping from their chariots. This rout caused the immediate dispersal of the forces that had assembled from various tribes to Cassivellaunus' aid, and the Britons never again joined battle with their whole strength.

—JANE MITCHELL
(translator)

VERGIL:

FROM THE

Aeneid

Vergil was born in Cisalpine Gaul in 70 B.C. His health being too poor to allow him to enter public life, he devoted himself to literature. When Augustus became emperor, he asked Vergil to write a monumental work to add to the glory of Rome. Vergil spent many years studying and preparing for this task and, in 19 B.C. he finally finished the first draft of his poem. He died that same year at Brundisium, before he could revise and polish the work as he intended. His Aeneid is the national epic of Rome, and one of the masterpieces of all world literature.

What god shall sing of the bitter strife, of the different
Ways to their doom of the slaughtered chiefs as now
Aeneas here and Turnus there quartered the battlefield?
O Jove, was it indeed your will that nations
Who were to live together in peace for ever
Should meet in such a clash?
Then Venus, most beautiful mother of Aeneas,
Put in his mind this thought: to march to the walls,
To switch his forces suddenly onto the city
And stun the Latians in a surprise attack.
And he, as he tracked through the battle,
Hither and thither, cast his hunting eye
On the safe city basking in its immunity
From the turmoil of the battle, aloof and quiet.
The vision of a more telling feat of arms
Immediately gripped his mind: he summoned his captains

Mnestheus, and Sergestus, and bold Serestus,
And standing on a mount to which the rest
Of the Trojan forces rallied in close order
Their weapons at the ready, standing there
On the top of the mound he spake these words to them.
"These are my orders: to be obeyed at once.
Jove is with us. Let nobody be the slower
Because this change of plan is a sudden one:
Today I propose to raze this city, the cause
Of the war, Latinus' capital. Unless
They acknowledge defeat and willingly submit
I will level its smoking turrets with the ground.
Am I going to wait for Turnus till he is pleased
To fight me? and, beaten, ask for a second chance?
O countrymen, here is the root and branch
Of this evil war. Fetch faggots! Exact with fire
The restoration of the broken treaty!"
Such were his words, and all his troops massed
Into a wedge and advanced to the city walls.
Suddenly, in a flash, scaling ladders and torches
Appeared and some of his men surprised the gateposts killing
The sentries, others discharged their spinning darts
And blackened the sky with weapons.

—Patrick Dickinson
(translator)

ONASANDER:

FROM

The General

Onasander lived in the first century A.D. *and* The General *was probably written around 55. He appears to have been a Platonic philosopher; whether he had any military experience is unknown. Onasander was clearly indebted to Xenophon and he, in turn, appears to have strongly influenced the Byzantine writers.*

A shrewd general who sees that the enemy has many troops when he himself is about to engage with fewer will select, or rather make it his practice to find, localities where he may prevent an encircling movement of the enemy, either by arranging his army along the bank of a river, or, by choosing a mountainous district he will use the mountains themselves to block off those who wish to outflank him, placing a few men on the summits to prevent the enemy from climbing above the heads of the main army. Not alone does knowledge of military science play a part in this matter but luck as well; for it is necessary to have the luck to find such places; one cannot prepare the terrain for oneself. To choose the better positions, however, from those at hand, and to know which will be advantageous, is the part of the wise general.

It is often the custom of generals who are in command of a powerful and numerous army to march in a crescent formation, believing that their opponents also wish the battle to come to close quarters, and that they will thus induce them to fight; then, as their opponents are bent back into the road at the points of the crescent, they will intercept them with their enveloping folds joining the extremes of their own wings to form a complete circle. Against troops advancing in this fashion, one should not likewise adopt the crescent formation, but dividing his own army into three parts the general should send two against the enemy, one against each wing, but the third division, that which faces the central hollow of the crescent, should stand still, opposite the enemy, and not advance.

For if the enemy maintain this crescent formation, those drawn up in the centre of their army will be useless, standing still and doing nothing; but if marching forward they wish to advance in a body, changing from the crescent formation to a straight line, they will be crowded together and will lose their formation—for while the wings are remaining in the same position and fighting, it is impossible for a crescent to return to straight line. Then when they are confused and their ranks disordered, the opposing general should send the third and reserve division against the men advancing in disorder from the centre of the curve. But if the enemy remain in the crescent position, the general should send his light-armed troops and archers opposite them, who with their missiles will cause heavy loss.

However, if he advances with his whole phalanx obliquely against one wing of the enemy, he will make no mistake in attacking in this manner, as far as the encircling movement of the crescent formation is concerned; for the enemy will be prevented for a long time from coming to close quarters with their whole army, and will be thrown into confusion little by little, since only those of one wing will be fighting, that is, those who will necessarily be the first to be engaged because of the oblique attack.

* * *

It is sometimes a useful stratagem for an enemy facing the enemy to retire gradually as if struck by fear, or to about face and make a retreat similar to a flight but in order, and suddenly turning, to attack their pursuers. For sometimes the enemy, delighted by the belief that their opponents are fleeing, break ranks and rush forward, leaping ahead of one another. There is no danger in turning to attack these men; and those who have for some time been pursuing, terrified by the very unexpectedness of this bold stand, immediately take to flight.

* * *

When passing through the country of an ally, the general must order his troops not to lay hands on the country, nor to pillage and destroy; for every army under arms is ruthless, when it has the opportunity of exercising power, and the close view of desirable objects entices the thoughtless to greediness; while small reasons alienate allies or make them quite hostile. But the country of the enemy he should ruin and burn and ravage, for loss of money and shortage of crops reduce warfare as abundance nourishes it. But first he should let the enemy know what he intends to do; for often the expectation of impending terror has brought those who have been endangered, before they have suffered at all, to terms which they previously would not have wished to accept; but when they have once suffered a reverse, in the belief that nothing can be worse they are careless of future perils.

—Illinois Greek Club
(translator)

TACITUS:

FROM

Annals

*Cornelius Tacitus was born in provincial Rome in A.D. 56
and studied rhetoric. He held a succession of minor offices
and was possibly in command of a legion for a time. He
became a consul in 95 and proconsul of Asia in 112. In later
years he concentrated on literature and especially on his
Histories. He died in 120 at the zenith of imperial Rome.*

*Germanicus finally defeats Arminius at Idestivus and re-
venges the Roman disaster under Varus.*

Arminius and the other German chiefs also each addressed their
men, reminding them that these Romans were Varus' runaways—men
who had mutinied to escape battle. Some had backs covered with
wounds, others were crippled by storm and sea; and now, hopeless,
deserted by the gods, they were again pitted against a relentless
enemy. They had taken to ships and remotest waters to evade attack
—and escape pursuit after disaster. "But once battle comes," cried
Arminius, "winds and oars cannot prevent their defeat!" He urged
his troops to remember how greedy, arrogant, and brutal Rome was.
The only alternatives, he insisted, were continued freedom or—in
preference to slavery—death.

Excited by this appeal, the Germans clamoured to fight. They
were marched to a level area called Idistaviso, which curves ir-
regularly between the Weser and the hills; at one point an outward
bend of the river gives it breadth, at another it is narrowed by
projecting high ground. Behind rose the forest, with lofty branches
but clear ground between the tree-trunks. The Germans occupied the

plain and the outskirts of the forest. The Cherusci alone occupied the heights, waiting to charge down when the battle started. The Roman army moved forward in the following order: first, Gallic and German auxiliaries followed by unmounted bowmen; next, four Roman brigades, and Germanicus with two battalions of the Guard and picked cavalry; then four more brigades, each brought by light infantry and mounted bowmen to divisional strength; and the remaining auxiliary battalions. The troops were alert and ready to deploy from column of march into battle order.

Units of the Cherusci charged impetuously. Seeing this, Germanicus ordered his best cavalry to attack their flank, while the rest of the cavalry, under Lucius Stertinius, was to ride round and attack them in the rear; and he himself would be there at the right moment. He saw a splendid omen—eight eagles flying towards and into the forest. "Forward," he cried, "follow the birds of Rome, the Roman army's protecting spirits!" The infantry attacked, and the cavalry, which had been sent ahead, charged the enemy's flanks and rear. It was a strange sight. Two enemy forces were fleeing in opposite directions, those from the woods into the open, those from the open into the woods.

The Cherusci between began to be dislodged from the slopes: among them Arminius, striking, shouting, wounded, trying to keep the battle going. His full force was thrown against the bowmen, and it would have broken through if the standards of the Raetian, Vindelician, and Gallic auxiliary battalions had not barred the way. Even so, by sheer physical strength aided by the impetus of his horse, he got through. To avoid recognition he had smeared his face with his own blood. One story is that Chauci among the Roman auxiliaries recognized him and let him go. Inguiomerus was likewise saved, by his own bravery or by treachery. The rest were massacred. Many tried to swim the Weser. They were battered by javelins, or carried away by the current, or finally overwhelmed by the mass of fugitives and collapse of the river banks. Some ignominiously tried to escape by climbing trees. As they cowered among the branches, bowmen amused themselves by shooting them down. Others were brought to the ground by felling the trees.

It was a great victory, and it cost us little. The slaughter of the enemy continued from midday until dusk. Their bodies and weapons were scattered for ten miles round. Among the spoils were found chains which they had brought for the Romans in confident expectation of the result. The troops hailed Tiberius as victor on the battle-field, and erected a mound on which, like a trophy, they set arms with the names of the defeated tribes. The sight of this upset and enraged the Germans more than all their wounds and losses and destruction. Men who had just been planning emigration across the Elbe now wanted to fight instead, and rushed to arms.

Germans of every rank and age launched sudden and damaging attacks against the Romans on the march. Finally they selected a narrow swampy open space enclosed between a river and the forest—which in its turn was surrounded by a deep morass (except on one side where a wide earthwork had been constructed by the Angrivarii to mark the Cheruscan frontier). Here the Germans stationed their infantry. The cavalry took cover in the woods nearby, so as to take the Romans in the rear when they came into the forest. Germanicus was aware of all this. He knew their plans, positions, their secret as well as their visible arrangements; and he planned to use their strategy for their own ruin. His cavalry, under the command of Lucius Seius Tubero, was allotted the open ground. The infantry were divided. Part were to proceed along the level ground to the wood, the rest were to scale the earthwork. He undertook this more difficult project himself, leaving the remaining tasks to his generals.

Those allocated the level ground broke into the wood easily. But the men scaling the earthwork were virtually climbing a wall, and received severe damage from above. Seeing that fighting conditions were unfavourable at close quarters, Germanicus withdrew his brigades a short way, and ordered his slingers into action to drive off the enemy. Simultaneously, spears were launched from machines. Exposure cost the defenders heavy casualties; they were beaten back, the earthwork was captured, and Germanicus personally led the Guard battalions in a charge into the woods. There, hand-to-hand fighting began. The enemy were hemmed in by the marsh behind them, the Romans by the river or hills. Both sides had to fight it out on the spot. Bravery was their only hope, victory their only way out.

The Germans were as brave as our men, but their tactics and weapons proved their downfall. With their vast numbers crammed into a narrow space they could neither thrust nor pull back their great pikes. They were compelled to fight as they stood, unable to exploit their natural speed by charging. The Romans on the other hand, with shields close to their chests and sword-hilts firmly grasped, rained blows on the enemy's huge forms and exposed faces, and forced a murderous passage.

Either Arminius had been through too many crises, or his recent wound was troubling him: he did not show his usual vigour. Inguiomerus, however, was in every part of the battle at once. His courage did not fail him—but he had bad luck. Germanicus, who had torn off his helmet so as to be recognized, ordered his men to kill and kill. No prisoners were wanted. Only the total destruction of the tribe would end the war. Finally, late in the day, he withdrew one brigade from the battle to make a camp. Apart from the cavalry, whose battle was indecisive, the rest sated themselves with enemy blood until nightfall.

Germanicus congratulated the victorious troops and piled up a heap of arms with this proud inscription: DEDICATED TO MARS AND THE DIVINE AUGUSTUS BY THE ARMY OF TIBERIUS CAESAR AFTER ITS CONQUEST OF THE NATIONS BETWEEN THE RHINE AND THE ELBE. Of himself he said nothing. He may have feared jealousy; or perhaps he felt that the knowledge of what he had done was enough. Shortly afterwards he sent Lucius Stertinius to fight the Angrivarii unless they rapidly surrendered. They begged for mercy unconditionally and received an unqualified pardon. Next, summer being already at its height, part of the army were sent back to winter quarters overland, while the majority embarked on the Ems and sailed with Germanicus down to the sea.

—MICHAEL GRANT
(translator)

ARRIAN:

FROM THE

Anabasis of Alexander

Flavius Arrianus was born in Bithynia in A.D. *96 and, unprecedently for a Greek, was appointed governor of a province, Cappadocia, in 131. He conducted a successful campaign against the Alani. He was a favorite of the emperor Hadrian and spent much of his time at Athens, holding public office there. He wrote numerous works on history, geography, philosophy, and some military treatises. His principal work, the* Anabasis of Alexander *is based on the lost history of Ptolemy, one of Alexander the Great's generals, and constitutes the main source of information on Alexander's generalship. Arrian died in 180.*

Alexander's victory over the Persians at Arbela (or Gaugamela) in 331 B.C. gives him the mastery of the world.

Such was the disposition of Alexander's front line, in addition to which he posted reserve formations in order to have a solid core of infantry to meet a possible attack from the rear; the officers of the reserve had orders, in the event of an encircling movement by the enemy, to face about and so meet the threatened attack. One half of the Agrianes, commanded by Attalus and in touch with the Royal Squadron on the right wing, were, together with the Macedonian archers under Brison, thrown forward at an oblique angle, in case it should suddenly prove necessary to extend or close up the front line of infantry, and in support of the archers was the so-called "Old Guard" of mercenaries under Cleander.

In advance of the Agrianes and archers were the advanced

scouts and the Paeonians, commanded by Aretes and Ariston; the mercenary cavalry commanded by Menidas were posted right in the van. The position in advance of the Royal Squadron and other units of the Companions was occupied by the other half of the Agriane contingent and of the archers, supported by Balacrus' spearmen who stood facing the Persian scythe-chariots. Menidas had orders to wheel and attack the enemy in the flank, should they attempt an outflanking movement.

So much for Alexander's right; on his left, forming an angle with the main body, were the Thracians under Sitalces supported, first, by the allied cavalry under Coeranus and, secondly, by the Odrysian cavalry under Agathon son of Tyrimmas. Right in the van of this sector was the foreign mercenary cavalry commanded by Andromachus son of Hieron. The Thracian infantry had orders to stand guard over the pack-animals. The total strength of Alexander's army was 7,000 cavalry and about 40,000 foot.

The two armies were now close together. Darius and his picked troops were in full view. There stood the Persian Royal Guard, the golden apples on their spear-butts, the Indians and Albanians, the Carians and the Mardian bowmen—the cream of the Persian force, full in face of Alexander as he moved with his Royal Squadron to the attack. Alexander, however, inclined slightly to his right, a move which the Persians at once countered, their left outflanking the Macedonians by a considerable distance. Meanwhile in spite of the fact that Darius' Scythian cavalry, moving along the Macedonian front, had already made contact with their forward units, Alexander continued his advance towards the right until he was almost clear of the area which the Persians had levelled during the previous days.

Darius knew that once the Macedonians reached rough ground his chariots would be useless, so he ordered the mounted troops in advance of his left to encircle the Macedonian right under Alexander and thus check any further extension in that direction. Alexander promptly ordered Menidas and his mercenary cavalry to attack them. A counter-attack by the Scythian cavalry and their supporting Bactrians drove them back by weight of numbers, whereupon Alexander sent in against the Scythians Ariston's Paeonian contingent and the mercenaries.

This stroke had its effect, and the enemy gave ground; but the remaining Bactrian units engaged the Paeonians and the mercenaries and succeeded in rallying the fugitives. A close cavalry action ensued, in which the Macedonians suffered the more severely, outnumbered as they were and less adequately provided with defensive armour than the Scythians were—both horses and men. None the less the Macedonians held their attacks, and by repeated counter-charges, squadron by squadron, succeeded in breaking the enemy formation.

Meanwhile as Alexander moved forward the Persians sent their scythe-chariots into action against him, in the hope of throwing his line into confusion. But in this they were disappointed; for the chariots were no sooner off the mark than they were met by the missile weapons of the Agrianes and Balacrus' javelin-throwers, who were stationed in advance of the Companions; again, they seized the reins and dragged the drivers to the ground, then surrounded the horses and cut them down. Some few of the vehicles succeeded in passing through, but to no purpose, for the Macedonians had orders, wherever they attacked, to break formation and let them through deliberately: this they did, with the result that neither the vehicles themselves nor their drivers suffered any damage whatever. Such as got through were, however, subsequently dealt with by the Royal Guard and the army grooms.

Darius now brought into action the main body of his infantry, and an order was sent to Aretes to attack the Persian cavalry which was trying to outflank and surround the Macedonian right. For a time Alexander continued to advance in column; presently, however, the movement of the Persian cavalry, sent to the support of their comrades who were attempting to encircle the Macedonian right, left a gap in the Persian front—and this was Alexander's opportunity. He promptly made for the gap, and, with his Companions and all the heavy infantry in this sector of the line, drove in his wedge and raising the battle-cry pressed forward at the double straight for the point where Darius stood.

A close struggle ensued, but it was soon over; for when the Macedonian horse, with Alexander himself at the head of them, vigorously pressed the assault, fighting hand to hand and thrusting at the Persians' faces with their spears, and the infantry phalanx in close order and bristling with pikes added its irresistible weight, Darius, who had been on edge since the battle began and now saw nothing but terrors all around him, was the first to turn tail and ride for safety. The outflanking party on the Macedonian right was also broken up by the powerful assault of Aretes and his men.

On this part of the field the Persian rout was complete, and the

Macedonians pressed the pursuit, cutting down the fugitives as they rode. But the formation under Simmias, unable to link up with Alexander to join in the pursuit, was forced to stand its ground and continue the struggle on the spot, a report having come in that the Macedonian left was in trouble. At this point the Macedonian line was broken, and some of the Indian and Persian cavalry burst through the gap and penetrated right to the rear where the Macedonian pack-animals were.

There was some hard fighting; the Persians set about it with spirit, most of their adversaries being unarmed men who had never expected a break-through—at any rate here, where the phalanx was of double strength; moreover, the prisoners joined in the attack. However, the officers in command of the reserves on this sector, the moment the situation was clear, faced about according to orders and appeared in the Persian rear. Many of the Persians, as they swarmed round the baggage-trains, were killed; others did not stay to fight, but made off.

Meanwhile the Persian right, not yet knowing that Darius had fled, made a move to envelop Alexander's left and delivered a flank attack on Parmenio. The Macedonians being caught, as it were, between two fires, Parmenio sent an urgent message to Alexander that his position was desperate and that he needed help. Alexander at once broke off the pursuit, wheeled about with his Companions and charged the Persian right at the gallop. Coming first into contact with those of the enemy cavalry who were trying to get away, he was soon heavily engaged with the Parthians, some of the Indians, and the strongest and finest cavalry units of Persia. The ensuing struggle was the fiercest of the whole action; one after another the Persian squadrons wheeled in file to the charge; breast to breast they hurled themselves on the enemy. Conventional cavalry tactics—manoeuvring, javelin-throwing—were forgotten; it was every man for himself, struggling to break through as if in that alone lay his hope of life. Desperately and without quarter, blows were given and received, each man fighting for mere survival without any further thought of victory or defeat. About sixty of Alexander's Companions were killed; among the wounded were Coenus, Menidas, and Hephaestion himself.

In this struggle Alexander was once again victorious. Such Persians as managed to fight their way through galloped off the field to save their skins.

Alexander was now on the point of engaging the Persian right;

but his help was not needed, as in this sector the Thessalian cavalry had fought hardly less magnificently than Alexander himself. The Persians were already in retreat by the time he made contact with them, so he turned back and started once more in pursuit of Darius, continuing as long as daylight served. Parmenio, in chase of his own quarry, was not far behind him. Once across the Lycus, Alexander halted for a brief rest for men and horses, and Parmenio went on to take possession of the Persian camp and all its contents; baggage, elephants, and camels.

—AUBREY DE SELINCOURT
(translator)

VEGETIUS:

FROM

Military Instructions

Flavius Vegetius Renatus was a celebrated military writer of the fourth century. Little is known of his life and military experience, but his Epitoma rei militaris, *though confused and unscientific, is invaluable to the student of the art of war. It was translated into English, French, and German in the fifteenth century.*

An army may be drawn up for a general engagement in seven different formations. The first formation is an oblong square of a large front, of common use both in ancient and modern times, although not thought the best by various judges of the service, because an even and level plain of an extent sufficient to contain its front cannot always be found, and if there should be any irregularity or hollow in the line, it is often pierced in that part. Besides, an enemy superior in number may surround either your right or left wing, the consequence of which will be dangerous, unless you have a reserve ready to advance and sustain his attack. A general should make use of this disposition only when his forces are better and more numerous than the enemy's, it being thereby in his power to attack both the flanks and surround them on every side.

The second and best disposition is the oblique. For although your army consists of few troops, yet good and advantageously posted, it will greatly contribute to your obtaining the victory, notwithstanding the numbers and bravery of the enemy. It is as follows: as the armies are marching up to the attack, your left wing must be kept back at such a distance from the enemy's right as to be out of reach of their darts and arrows. Your right wing must advance obliquely upon the enemy's left, and begin the engagement. And you must endeavor with your best cavalry and infantry to surround the wing with which you are engaged, make it give way and fall upon the enemy in the rear. If they once give ground and the attack is properly seconded, you will undoubtedly gain the victory, while your left wing, which continued at a distance, will remain untouched. An army drawn up in this manner bears some resemblance to the letter A or a mason's level. If the enemy should be beforehand with you in this evolution, recourse must be had to the supernumerary horse and foot posted as a reserve in the rear, as I mentioned before. They must be ordered to support your left wing. This will enable you to make a vigorous resistance against the artifice of the enemy.

The third formation is like the second, but not so good, as it obliges you to begin the attack with your left wing on the enemy's right. The efforts of soldiers on the left are weak and imperfect from their exposed and defective situation in the line. I will explain this formation more clearly. Although your left wing should be much better than your right, yet it must be reinforced with some of the best horse and foot and ordered to commence the action with the enemy's right in order to disorder and surround it as expeditiously as possible. And the other part of your army, composed of the worst troops, should remain at such a distance from the enemy's left as not to be annoyed by their darts or in danger of being attacked sword in hand. In this oblique formation care must be taken to prevent the line being penetrated by the wedges of the enemy, and it is to be employed only when the enemy's right wing is weak and your greatest strength is on your left.

The fourth formation is this: as your army is marching to the attack in order of battle and you come within four or five hundred paces of the enemy, both your wings must be ordered unexpectedly to quicken their pace and advance with celerity upon them. When they find themselves attacked on both wings at the same time, the sudden surprise may so disconcert them as to give you an easy victory. But although this method, if your troops are very resolute and expert, may ruin the enemy at once, yet it is hazardous. The general who attempts it is obliged to abandon and expose his center and to divide his army into three parts. If the enemy are not routed at the first charge, they have a fair opportunity of attacking the wings which are separated from each other and the center which is destitute of assistance.

The fifth formation resembles the fourth but with this addition: the light infantry and the archers are formed before the center to cover it from the attempts of the enemy. With this precaution the general may safely follow the above mentioned method and attack the enemy's left wing with his right, and their right with his left. If he puts them to flight, he gains an immediate victory, and if he fails of success his center is in no danger, being protected by the light infantry and archers.

The sixth formation is very good and almost like the second. It is used when the general cannot depend either on the number or courage of his troops. If made with judgment, notwithstanding his inferiority, he has often a good chance for victory. As your line approaches the enemy, advance your right wing against their left and begin the attack with your best cavalry and infantry. At the same time keep the rest of the army at a great distance from the enemy's right, extended in a direct line like a javelin. Thus if you can surround their left and attack it in flank and rear, you must inevitably defeat them. It is impossible for the enemy to draw off reinforcements from their right or from their center to sustain their left in this emergency, since the remaining part of your army is extended and at a great distance from them in the form of the letter L. It is a formation often used in an action on a march.

The seventh formation owes its advantages to the nature of the ground and will enable you to oppose an enemy with an army inferior both in numbers and goodness, provided one of your flanks can be covered either with an eminence, the sea, a river, a lake, a city, a morass or broken ground inaccessible to the enemy. The rest of the army must be formed, as usual, in a straight line and the unsecured flank must be protected by your light troops and all your cavalry. Sufficiently defended on one side by the nature of the ground and on the other by a double support of cavalry, you may then safely venture on action.

One excellent and general rule must be observed. If you intend to engage with your right wing only, it must be composed of your best troops. And the same method must be taken with respect to the left. Or if you intend to penetrate the enemy's line, the wedges which you form for that purpose before your centre, must consist of the best disciplined soldiers. Victory in general is gained by a small number of men. Therefore the wisdom of a general appears in nothing more

than in such choice of disposition of his men as is most consonant
with reason and service.

—THOMAS R. PHILLIPS
(translator)

PROCOPIUS:

FROM

History of the Gothic Wars

Procopius was born in Palestine in 507, received a legal training, and became adviser to the Byzantine general Belisarius during his campaigns against Persia, Africa, and Italy. By 543 he had returned to Constantinople and may subsequently have been a prefect there. Apart from the Wars *and a book on buildings, he wrote—but did not publish—a* Secret History, *which contains a violent attack on the Emperor Justinian, whom he had eulogized in his published works.*

In his History of the Gothic Wars, *Procopius describes the Byzantine strategy and tactics in Italy.*

And Belisarius thought that henceforth his army ought to carry the war against the enemy. On the following day, accordingly, he commanded one of his own bodyguards, Trajan by name, an impetuous and active fighter, to take two hundred horsemen of the guards and go straight towards the enemy, and as soon as they came near the camps to go up on a high hill (which he pointed out to him) and remain quietly there. And if the enemy should come against them, he was not to allow the battle to come to close quarters, nor to touch sword or spear in any case, but to use bows only, and as soon as he should find that his quiver had no more arrows in it, he was to flee as hard as he could with no thought of shame and retire to the fortifications on the run.

Having given these instructions, he held in readiness both the engines for shooting arrows and the men skilled in their use. Then Trajan with the two hundred men went out from the Salarian Gate against the camp of the enemy. And they, being filled with amazement at the suddenness of the thing, rushed out from the camps, each man equipping himself as well as he could. But the men under Trajan galloped to the top of the hill which Belisarius had shown them, and from there began to ward off the barbarians with missiles. And since their shafts fell among a dense throng, they were for the most part successful in hitting a man or a horse. But when all their missiles had at last failed them, they rode off to the rear with all speed, and the Goths kept pressing upon them in pursuit. But when they came near the fortifications, the operators of the engines began to shoot arrows from them, and the barbarians became terrified and abandoned the pursuit. And it is said that not less than one thousand Goths perished in this action.

A few days later Belisarius sent Mundilas, another of his own bodyguard, and Diogenes, both exceptionally capable warriors, with three hundred guardsmen, commanding them to do the same thing as the others had done before. And they acted according to his instructions. Then, when the enemy confronted them, the result of the encounter was that no fewer than in the former encounter, perhaps even more, perished. And sending even a third time the guardsman Oilas with three hundred horsemen, with instructions to handle the enemy in the same way, he accomplished the same result. So in making these three sallies, in the manner told by me, Belisarius destroyed about four thousand of his antagonists.

But Vittigis failing to take into account the difference between the two armies in point of equipment of arms and of practice in warlike deeds, thought that he too would most easily inflict grave losses upon the enemy if only he should make his attack upon them with a small force.

He therefore sent five hundred horsemen, commanding them to go close to the fortifications, and to make a demonstration against the whole army of the enemy of the very same tactics as had time and again been used against them, to their sorrow, by small bands of the foe. And so, when they came to a high place not far from the city, but just beyond the range of the missiles, they took their stand there. But Belisarius selected a thousand men, putting Bessas in command, and ordered them to engage the enemy. And this force, by forming a circle around the enemy and always shooting at them from behind, killed a large number, and by pressing hard upon the rest compelled them to descend into the plain. There a hand-to-hand battle took place between forces not evenly matched in strength, and most of the

Goths were destroyed, though some few with difficulty made their escape and returned to their own camp. . . .

And the difference was this, that practically all the Romans and their allies, the Huns, are good mounted bowmen, but not a man among the Goths has had practice in this branch, for their horsemen are accustomed to use only spears and swords, while their bowmen enter battle on foot and under cover of heavily armed men. So the horsemen, unless the engagement is at close quarters, have no means of defending themselves against opponents who can use the bow, and therefore can easily be reached by the arrows and destroyed; and as for the foot-soldiers, they can never be strong enough to make sallies against men on horseback. It was for these reasons, Belisarius declared, that the barbarians had been defeated by the Romans, in these last engagements.

—H. B. DEWING
(translator)

LEO:

FROM

Tactica

The Tactica *is considered to have been written by, or under
the direction of, the Byzantine emperor Leo VI. He reigned
from 886 until his death in 911 and is known as the Wise
or the Philosopher, because he wrote a number of theologi-
cal and poetic works. He did much to reform the civil
administration of the empire.*

Some authorities attribute the Tactica *to Leo III who
reigned from 717 to 740. Known as the Isaurian, he had a
distinguished military career and his successful resistance
to a Saracen siege of Constantinople, as well as administrative reforms, saved the empire from collapse.*

The Franks and Lombards are bold and daring to excess,
though the latter are no longer all that they once were; they regard
the smallest movement to the rear as a disgrace, and they will fight
whenever you offer them battle. When their knights are hard put to it
in a cavalry fight, they will turn their horses loose, dismount and
stand back to back against very superior numbers rather than fly. So
formidable is the charge of the Frankish chivalry with their broad-
sword, lance, and shield, that it is best to decline a pitched battle with
them till you have put all the chances on your side.

You should take advantage of their indiscipline and disorder;
whether fighting on foot or on horseback, they charge in dense, unwieldy masses, which cannot manoeuvre, because they have neither
organization nor drill. Tribes and families stand together, or the
sworn war-bands of chiefs, but there is nothing to compare to our
own orderly division into battalions and brigades. Hence they readily

fall into confusion if suddenly attacked in flank and rear—a thing easy to accomplish, as they are utterly careless and neglect the use of pickets and vedettes and the proper surveying of the countryside. They encamp, too, confusedly and without fortifying themselves, so that they can be easily cut up by a night attack.

Nothing succeeds better against them than a feigned flight, which draws them into an ambush; for they follow hastily, and invariably fall into a snare. But perhaps the best tactics of all are to protract the campaign, and lead them into the hills and desolate tracts, for they take no care about their commissariat, and when their stores run low their vigour melts away. They are impatient of hunger and thirst, and after a few days of privation desert their standards and steal away home as best they can. For they are destitute of all respect for their commanders—one noble thinks himself as good as another—and they will deliberately disobey orders when they grow discontented. Nor are their chiefs above the temptation of taking bribes; a moderate sum of money will frustrate one of their expeditions.

On the whole, therefore, it is easier and less costly to wear out a Frankish army by skirmishes, protracted operations in desolate districts, and the cutting off of its supplies, than to attempt to destroy it at a single blow.

—CHARLES OMAN
(translator)

JUVAINI:

FROM

History of the World Conqueror

*Ala Ud-Din, Ata-Malik Juvaini was born in Persia in 1226,
the grandson of a high official of the shahs in their resistance
to the invasion of Jenghiz Khan. Juvaini accompanied his
father on visits to Mongolia and there began his history of
the Mongols. On his return to the West he was attached to
the staff of Hulagu, the Mongol conqueror of Baghdad,
and accompanied him on his campaigns. In 1260 he was
appointed governor of the conquered territories. He died
in 1283.*

What army in the whole world can equal the Mongol army? In
time of action, when attacking and assaulting, they are like trained
wild beasts out after game, and in the days of peace and security they
are like sheep, yielding milk, and wool, and many other useful things.
In misfortune and adversity they are free from dissension and opposition.

It is an army after the fashion of a peasantry, being liable to all
manner of contributions and rendering without complaint whatever is
enjoined upon it, whether occasional taxes, the maintenance of
travellers or the upkeep of post stations (*yam*) with the provision of
mounts and food therefor. It is also a peasantry in the guise of an
army, all of them, great and small, noble and base, in time of battle
becoming swordsmen, archers and lancers and advancing in whatever
manner the occasion requires.

Whenever the slaying of foes and the attacking of rebels is pur-
posed, they specify all that will be of service for that business, from
the various arms and implements down to banners, needles, ropes,

mounts and pack animals such as donkeys and camels; and every man must provide his share according to his ten or hundred. On the day of review, also, they display their equipment, and if only a little be missing, those responsible are severely punished. Even when they are actually engaged in fighting, there is exacted from them as much of the various taxes as is expedient, while any service which they used to perform when present devolves upon their wives and those of them that remain behind. Thus if work be afoot in which a man has his share of forced labour, and if the man himself be absent, his wife goes forth in person and performs that duty in his stead.

The reviewing and mustering of the army has been so arranged that they have abolished the registry of inspection and dismissed the officials and clerks. For they have divided all the people into companies of ten, appointing one of the ten to be the commander of the nine others; while from among each ten commanders one has been given the title of "commander of the hundred," all the hundred having been placed under his command. And so it is with each thousand men and so also with each ten thousand, over whom they have appointed a commander whom they call "commander of the *tümen*."

In accordance with this arrangement, if in an emergency any man or thing be required, they apply to the commanders of *tümen;* who in turn apply to the commanders of thousands, and so on down to the commanders of tens. There is a true equality in this; each man toils as much as the next, and no difference is made between them, no attention being paid to wealth or power. If there is a sudden call for soldiers an order is issued that so many thousand men must present themselves in such and such a place at such and such an hour of that day or night. And they arrive not a twinkling of an eye before or after the appointed hour.

Their obedience and submissiveness is such that if there be a commander of a hundred thousand between whom and the Khan there is a distance of sunrise and sunset, and if he but commit some fault, the Khan dispatches a single horseman to punish him after the manner prescribed: if his head has been demanded, he cuts it off, and if gold be required, he takes it from him.

Another *yasa* is that no man may depart to another unit than the hundred, thousand or ten to which he has been assigned, nor may he seek refuge elsewhere. And if this order be transgressed the man who transferred is executed in the presence of the troops, while he that received him is severely punished. For this reason no man can give refuge to another; if (for example) the commander be a prince, he does not permit the meanest person to take refuge in his company and so avoids a breach of the *yasa*. Therefore no man can take liberties with his commander or leader, nor can another commander entice him away.

Furthermore, when moonlike damsels are found in the army they are gathered together and dispatched from the tens to the hundreds, and each man makes a different choice up to the commander of the *tümen,* who makes his choice also and takes the maidens so chosen to the Khan or the princes. These too make their selection, and upon those that are deemed worthy and are fair to look upon they recite the words *"Keep them honourably,"* and upon the other, *"Put them away with kindness."* And they cause them to attend on the Royal Ladies until such time as it pleases them to bestow them on others or to lie with them themselves.

Again, when the extent of their territories became broad and vast and important events fell out, it became essential to ascertain the activities of their enemies, and it was also necessary to transport goods from the West to the East and from the Far East to the West. Therefore throughout the length and breadth of the land they established *yams,* and made arrangements for the upkeep and expenses of each *yam,* assigning thereto a fixed number of men and beasts as well as food, drink and other necessities. All this they shared out amongst the *tümen,* each two *tümen* having to supply one *yam.* Thus, in accordance with the census, they so distribute and exact the charge, that messengers need make no long detour in order to obtain fresh mounts while at the same time the peasantry and the army are not placed in constant inconvenience. Moreover strict orders were issued to the messengers with regard to the sparing of the mounts, etc., to recount all of which would delay us too long. Every year the *yams* are inspected, and whatever is missing or lost has to be replaced by the peasantry.

—JOHN BOYLE
(translator)

II.
RENAISSANCE
AND
REFORMATION

We may gather out of history a policy no less wise than eternal: by comparison and application of other men's forepassed miseries with our own like errors.

—RALEIGH

NICCOLO MACHIAVELLI:

FROM

The Prince;

FROM

The Art of War

Niccolo Machiavelli was born in Florence in 1469. He was largely self-educated, and at the age of twenty-one was appointed to an important city office. He spent the following years on various missions, sometimes abroad, and met Cesare Borgia—who became the model for The Prince. *He became a protagonist of a state militia which was established in Florence in 1504 under his supervision and commanded the troops in the capture of Pisa in 1509. In 1512 the Medici returned to power and Machiavelli was deprived of his power, imprisoned, and tortured. After his release he continued to write histories, comedies, and poetry besides his most celebrated works,* The Prince *and* The Discourses. *He gradually won favor with the Medicis, returning to public office. He joined the papal armies against Charles V until the sack of Rome in 1527. On the fall of the Medicis he found himself compromised with the new regime and died the same year.*

FROM The Prince

A prince, then, is to have no other design, nor thought, nor study but war and the arts and disciplines of it; for, indeed, that is the only profession worthy of a prince, and is of so much importance that it not only preserves those who are born princes in their patrimonies but advances men of private condition to that honourable degree. On the other side, it is frequently seen, when princes have addicted themselves more to delicacy and softness than to arms, they have lost all, and been driven out of their States; for the principal thing which deprives or gains a man authority is the neglect or profession of that art.

He never, therefore, ought to relax his thoughts from the exercises of war not so much as in time of peace; and, indeed, then he should employ his thoughts more studiously therein than in war itself, which may be done two ways, by the application of the body and the mind.

As to his bodily application, or matter of action, besides that he is obliged to keep his armies in good discipline and exercise, he ought to inure himself to sports, and by hunting and hawking, and such like recreation, accustom his body to hardship, and hunger, and thirst, and at the same time inform himself of the coasts and situation of the country, the bigness and elevation of the mountains, the largeness and avenues of the valleys, the extent of the plains, the nature of the rivers and fens, which is to be done with great curiosity; and his knowledge is useful two ways, for hereby he not only learns to know his own country and to provide better for its defence, but it prepares and adapts him, by observing their situations, to comprehend the situations of other countries, which will perhaps be necessary for him to discover; for the hills, the vales, the plains, the rivers, and the marshes (for example, in Tuscany), have a certain similitude and resemblance with those in other provinces; so that, by the knowledge of one, we may easily imagine the rest; and that prince who is defective in this, wants the most necessary qualification of a general; for by knowing the country, he knows how to beat up his enemy, take up his quarters, march his armies, draw up his men, and besiege a town with advantage.

In the character which historians give of Philopomenes, Prince of Achaia, one of his great commendations is, that in time of peace he thought of nothing but military affairs, and when he was in company with his friends in the country, he would many times stop suddenly and expostulate with them: If the enemy were upon that hill, and our army where we are, which would have the advantage of the ground? How could we come at them with most security? If we would draw off, how might we do it best? Or, if they would retreat, how might we follow? So that as he was traveling, he would propose all the accidents to which an army was subject; he would hear their opinion, give them his own, and reinforce it with arguments; and this he did so frequently, that by continual practice and a constant intention of his thoughts upon that business, he brought himself to that perfection, no accident could happen, no inconvenience could occur to an army, but he could presently redress it.

But as to the exercise of the mind, a prince is to do that by diligence in history and solemn consideration of the actions of the most excellent men, by observing how they demeaned themselves in the wars, examining the grounds and reasons of their victories and losses, that he may be able to avoid the one and imitate the other; and above all, to keep close to the example of some great captain of old (if any such occurs in his reading), and not only to make him his pattern, but to have all his actions perpetually in his mind, and it was said Alexander did by Achilles, Caesar by Alexander, Scipio by Cyrus.

—LESLIE WALKER SJ
(translator)

FROM *The Art of War*

As to your saying that I directed the shots of artillery according to my own will by making them pass over the heads of the infantry, I reply to you that there are many more times, beyond comparison, when the heavy guns do not strike the infantry than when they do strike them. For the infantry is so low and the guns are so hard to manage that, if you raise them a little, they fire over the heads of the infantry, and if you lower them, they hit the earth and the shot does not reach the troops. The infantry are protected also by the unevenness of the land, because every little thicket or bank between them and the guns is a protection. The horses, and especially those of the men-at-arms, because they have to stand closer together than the light horses and, being taller, are more easily hit, can be kept in the rear until the artillery has fired.

It is true that much more harm is done by the harquebuses and the light artillery than by the heavy, against which the chief protection is to come to close quarters quickly; if in the first attack they kill somebody, in such conditions somebody always is killed. A good general and a good army do not fear an individual harm but a universal one; they imitate the Swiss, who never avoid a battle because they are dismayed by the artillery; on the contrary, they inflict capital punishment on those who for fear of it either leave their rank or with their bodies give any sign of fear. I had the artillery, as soon as it had fired, retire into the army that it might give free passage to the battalions. I made no further mention of it because it is useless in combat at close quarters.

You have also said that because of the fury of this instrument, many consider the ancient arms and methods useless; this speech of yours suggests that the moderns have found methods and weapons that against artillery are effective. If you know any, I should be glad to have you explain them to me, because up to now I have never seen any of them nor do I believe any can be found. Indeed I should like to ask for what reasons the footsoldiers in our times wear the breastplate and the corselet of steel, and men on horseback are completely covered with armor, because those who condemn the ancient wearing of armor as useless against artillery, they ought to give up modern armor too. I should like to know for what reason the Swiss, imitating ancient methods, form a closepacked brigade of six or eight thousand infantrymen and for what reason all other infantry imitates them, since this order is exposed to the same danger, with respect to artillery, to which are exposed all others that might be imitated from antiquity.

I believe a champion of artillery would not know what to answer; but if you should ask soldiers who have some judgment, they would answer, first, that they go armored because, though that armor does not protect them from artillery, it does protect them from arrows, from pikes, from swords, from stones and from every other injurious thing that comes from the enemy. They would also answer that they go drawn up in close order like the Swiss in order to be able more easily to charge infantry, to resist cavalry better, and to give the enemy more difficulty in breaking them. Evidently soldiers need to fear many things besides artillery; from these they protect themselves with armor and with good order. From this it follows that the better armored an army is and the more its ranks are locked together and strong, the more secure it is.

Hence any who think artillery all-important must be either of little prudence or must have thought very little on these things. If we see that a very small part of the ancient method of arming that is used today, namely, the pike, and a very small part of their organization, namely, the brigades of the Swiss, do us so much good and give our armies so much strength, why are we not to believe that other arms and other customs that have been given up are useful? Besides, if we do not consider the artillery in putting ourselves in close formation like the Swiss, what other methods can make us fear it more? For evidently no arrangement can make us fear it so much as those that crowd men together.

In addition, the artillery of the enemy does not frighten me off from putting myself with my army near a city where it can harm me with full security, since I cannot take it because it is protected by walls which permit it to repeat its shots at pleasure; indeed I can only hinder it, in time, with my artillery. Why, then, am I to fear it in the field where I can quickly take it? Hence I conclude with this: artillery, in my opinion, does not make it impossible to use ancient methods and show ancient vigor.

—ALLAN GILBERT
(translator)

RICHARD HAKLUYT:

FROM

Voyages and Discoveries

Richard Hakluyt was born near London in 1553 and after graduating from Oxford devoted his life and fortune to the development of geographical science. As a chaplain he himself was employed in secretly collecting information about French and Spanish maritime movements, particularly with regard to America. From merchants, sea captains, and others he amassed and translated, when necessary, a wealth of contemporary information of scientific, economic, and strategic value. He himself wrote the Particular Discourse on Western Discoveries *in 1584, setting out the arguments for colonization, as well as a commentary on Aristotle's* Politics. *Sir Walter Raleigh was his patron, and he was closely associated with him in the Virginia project. He was appointed Archdeacon of Westminster and died in 1616.*

"The miraculous victory achieved by the English fleet upon the Spanish hugh Armada sent in the year 1588 for the invasion of England." (Recorded by Emmanuel van Meteren in his *History of the Low Countries*.)

After being harried by the smaller English ships up the Channel, the Armada put in at Calais. . . .

Whenas therefore the Spanish fleet rode at anchor before Calais, the Lord Admiral of England took forthwith eight of his worst and basest ships which came next to hand, and disburdening them of all things which seemed to be of any value, filled them with gun-powder, pitch, brimstone, and with other combustible and fiery matter; and charging all their ordnance with powder, bullets, and stones, he sent the said ships upon the 28 of July being Sunday, about two of the clock after midnight, with the wind and tide against the Spanish fleet: which being forsaken of the pilots and set on fire, were directly carried upon the King of Spain's navy: which fire in the dead of the night put the Spaniards into such a perplexity and horror that cutting their cables whereon their anchors were fastened, and hoisting up their sails, they betook themselves very confusedly unto the main sea.

In this sudden confusion, the principal and greatest of the four galliasses falling foul of another ship, lost her rudder: for which cause when she could not be guided any longer, she was by the force of the tide cast into a certain shoal upon the shore of Calais, where she was immediately assaulted by divers English pinnaces.

This huge and monstrous galliasse, wherein were contained three hundred slaves to lug at the oars, and four hundred soldiers, was in the space of three hours rifled in the same place; and there were found amongst divers other commodities 50,000 ducats of the Spanish king's treasure. At length the slaves were released out of their fetters.

Albeit there were many excellent and warlike ships in the English fleet, yet scarce were there 22 or 23 among them all which matched 90 of the Spanish ships in bigness, or could conveniently assault them. Wherefore the English ships using their prerogative of nimble steerage, whereby they could turn and wield themselves with the wind whichever way they listed, came often times very near upon the Spaniards, and charged them so sore, that now and then they were but a pike's length asunder: and so continually giving them one broad side after another, they discharged all their shot both great and small upon them, spending one whole day from morning till night in that violent kind of conflict, until such time as powder and bullets failed them.

The Spaniards that day sustained great loss and damage having many of their ships shot through and through, and they discharged likewise great store of ordnance against the English; who indeed sustained some hindrance, but not comparable to the Spaniards' loss; for they lost not any one ship or person of account. Albeit Sir Francis Drake's ship was pierced with shot above forty times, and his very cabin was twice shot through, and about the conclusion of the fight, the bed of a certain gentleman lying weary thereupon, was taken quite from under him with the force of a bullet. Likewise, as the Earl

of Northumberland and Sir Charles Blunt were at dinner upon a
time, the bullet of a demi-culverin broke through the midst of their
cabin, touched their feet, and struck down two of the standers by,
with many such accidents befalling the English ships, which it were
tedious to rehearse. Whereupon it is most apparent, that God miracu-
lously preserved the English nation.

The same night two Portuguese galleons of the burthen of seven
or eight hundred tons apiece, to wit the *Saint Philip* and the *Saint
Matthew,* were forsaken of the Spanish fleet, for they were so torn
with shot, that the water entered into them on all sides.

The 29 of July the Spanish fleet being encountered by the English lying close together under their fighting sails, with a southwest
wind sailed past Dunkirk, the English ships still following the chase.
The Lord Admiral of England despatched the Lord Henry Seymour
with his squadron of small ships unto the coast of Flanders, where,
with the help of the Dutch ships, he might stop the Prince of Parma
his passage, if perhaps he should attempt to issue forth with his army.
And he himself in the mean space pursued the Spanish fleet until the
second of August, because he thought they had set sail for Scotland.
And albeit he followed them very near, yet did he not assault them
any more, for want of powder and bullets. But upon the fourth of
August, the wind arising, when as the Spaniards had spread all their
sails, betaking themselves wholly to flight, and leaving Scotland on
the left hand, trended toward Norway.

The English seeing that they were now proceeded unto the latitude of 57 degrees, and being unwilling to participate that danger
whereinto the Spaniards plunged themselves, and because they
wanted things necessary, and especially powder and shot, returned
back for England; leaving behind them certain pinnaces only, which
they enjoined to follow the Spaniards aloof, and to observe their
course.

The Spaniards seeing now that they wanted four or five thousand of their people and having divers maimed and sick persons, and
likewise having lost 10 or 12 of their principal ships, they consulted
among themselves, what they were best to do, being now escaped out
of the hands of the English. They thought it good at length, so soon
as the wind should serve them, to fetch a compass about Scotland
and Ireland, and so to return for Spain.

They well understood, that commandment was given throughout
all Scotland, that they should not have any succour or assistance

there. Neither yet could they in Norway supply their wants. Fearing also lest their fresh water should fail them, they cast all their horses and mules overboard: and so touching nowhere upon the coast of Scotland, but being carried with a fresh gale between the Orkneys and Fair Isle, they proceeded far north, even unto 61 degrees of latitude, being distant from any land at the least 40 leagues. Here the Duke of Medina general of the fleet commanded all his followers to shape their course for Biscay: and he himself with twenty or five and twenty of his ships which were best provided of fresh water and other necessaries, holding on his course over the main ocean, returned safely home.

There arrived at Newhaven in Normandy, being by the tempest enforced to do so, one of the four great galliasses, where they found the ships with the Spanish women which followed the fleet at their setting forth. Two ships also were cast away upon the coast of Norway, one of them being of a great burthen; howbeit all the persons in the said great ship were saved: insomuch that of 134 ships, which set sail out of Portugal, there returned home 53 only small and great: namely of the four galliasses but one, and but one of the four galleys. Of the 91 great galleons and hulks there were missing 58, and 33 returned. Of 30,000 persons which went in this expedition, there perished (according to the number and proportion of the ships) the greater and better part; and many of them which came home, by reason of the toils and inconveniences which they sustained in this voyage, died not long after their arrival.

WALTER RALEIGH:

FROM

History of the World

Walter Raleigh was born in Devonshire in 1554 and appears to have been educated at Oxford. In 1578 he went to sea, destroying Spanish shipping with his half-brother, Sir Humphrey Gilbert. In 1580 he fought ruthlessly against the Irish rebels and his plans for a more rapid subjugation of Ireland brought him to the notice of Queen Elizabeth. He became her favorite and soon amassed a fortune from monopolies, as well as vast estates in Ireland. His part in the defeat of the Spanish Armada is controversial. In 1584 he sent out an expedition for the settlement of North America, naming the country Virginia.

A secret affair with one of the Queen's maids of honor led to his fall from favor and imprisonment in 1592. After his release he sailed for the Orinoco in South America, to find gold and establish a new colonial El Dorado—without success. In 1596 he took part in the attack on Cadiz in which he was wounded and the following year commanded an amphibious operation in the Azores.

On the accession of James I he was accused of treason and imprisoned in the Tower of London for thirteen years. Here he occupied himself with scientific experiments and the composition of his History of the World. *In 1616 he was released to undertake another search for gold, in Guiana. A disastrous assault on the Spanish positions, against orders, followed by an abortive flight to France ended with his execution at Westminster in 1618. His unfinished* History of the World *was read and admired by Cromwell among others. He was also a poet of distinction.*

Certainly, he that will happily perform a fight at sea, must be skilful in making choice of vessels to fight in; he must believe that there is more belonging to a good man of war, upon the waters, than great daring; and must know that there is a great deal of difference between fighting loose or at large, and grappling. The guns of a slow ship pierce as well, and make as great holes, as those in a swift. To clap ships together, without consideration, belongs rather to a mad man than to a man of war: for by such ignorant bravery was Peter Strozzi lost at the Azores, when he fought against the Marquess of Santa Cruz.

In like sort had the Lord Charles Howard, Admiral of England, been lost in the year 1588, if he had not been better advised than a great many malignant fools were, that found fault with his demeanour. The Spaniards had an army aboard them; and he had none: they had more ships than he had, and of higher building and charging; so that, had he entangled himself with those great and powerful vessels, he had greatly endangered this Kingdom of England. For twenty men upon the defences are equal to an hundred that board and enter; whereas then, contrariwise, the Spaniards had an hundred, for twenty of ours, to defend themselves withal. But our Admiral knew his advantage, and held it: which had he not done, he had not been worthy to have held his head.

Here to speak in general of Seafight (for particulars are fitter for private hands than for the press), I say that a fleet of twenty ships, all good sailors, and good ships, have an advantage, on the open sea, of an hundred as good ships, and slower sailing. For if the fleet of an hundred sail keep themselves near together, in a gross squadron, the twenty ships, charging them upon any angle, shall force them to give ground, and to fall back upon their own next fellows, of which so many as entangle, are made unserviceable or lost. Force them they may easily, because the twenty ships, which give themselves scope, after they have given one broadside of artillery, by clapping into the wind and staying, they may give them the other: and so the twenty ships batter them in pieces with a perpetual volley: whereas those that fight in a troop have no room to turn, and can always use but one and the same beaten side.

If the fleet of an hundred sail give themselves any distance, then shall the lesser fleet prevail, either against those that are a-rear and hindmost, or against those that by advantage of oversailing their fellows keep the wind: and if upon a lee-shore, the ships next the

wind be constrained to fall back into their own squadron, then it is all
to nothing, the whole fleet must suffer shipwreck, or render itself.

* * *

It is impossible for any maritime country, not having the coasts
admirably fortified, to defend itself against a powerful enemy that is
master of the sea. Let us consider of the matter itself; what another
nation might do, even against England, in landing an army, by advantage of a fleet, if we had none.

This question, whether an invading army may be resisted at their
landing upon the coast of England, were there no fleet of ours at the
sea to impeach it, is already handled by a learned gentleman of our
nation, in his observations upon Caesar's *Commentaries,* that maintains the affirmative. This he holds only upon supposition, in absence
of our shipping: and comparatively, as that it is a more safe and easy
course to defend all the coast of England than to suffer any enemy to
land, and afterwards to fight with him.

Surely I hold with him, that it is the best way, to keep our
enemy from treading upon our ground; wherein, if we fail, then must
we seek to make him wish that he had stayed at his own home. But
making the question general, and positive, whether England, without
help of her fleet, be able to debar an enemy from landing; I hold that
it is unable so to do: and therefore I think it most dangerous to make
the adventure. For the encouragement of a first victory to an enemy,
and the discouragement of being beaten to the invaded, may draw it a
most perilous consequence.

Our question is, of an army to be transported over sea, and to
be landed again in an enemy's country, and the place left to the
choice of the invader. Hereunto I say that such an army cannot be
resisted on the coast of England, without a fleet to impeach it; no,
nor on the coast of France, or any other country; except every creek,
port, or sandy bay had a powerful army in each of them, to make opposition. . . .

For there is no man ignorant that ships, without putting themselves out of breath, will easily outrun the soldiers that coast them.
Les Armées ne volent point en poste; Armies neither fly, nor run
post, saith a Marshal of France. And I know it to be true that a fleet
of ships may be seen at sunset, and after it, at the Lizard; yet by the
next morning they may recover Portland, whereas an army of foot
shall not be able to march it in six days.

For the end of this digression, I hope that this question shall never come to trial; his Majesty's many movable forts will forbid the experience. And although the English will no less disdain, than any nation under heaven can do, to be beaten upon their own ground, or elsewhere by a foreign enemy; yet to entertain those that shall assail us, with their own beef in their bellies, and before they eat of our Kentish capons, I take it to be the wisest way. To do which, his Majesty, after God, will employ his good ships on the sea, and not trust to any entrenchment upon the shore.

HUGO GROTIUS:

FROM

On the Laws of War and Peace

Hugo Grotius (Huigh de Groot), was born in Delft, Holland, in 1583 and early acquired a reputation as a scholar. At fifteen he edited an encyclopedia and accompanied Johan van Oldenbarneveldt, the leading Dutch statesman to the French court. Here he wrote poetry, studied law, and on his return to Holland was appointed official historiographer of the states of Holland—at the age of eighteen. He became involved in disputes about maritime law and developed his writings on the subject into a general consideration of international law—published in 1625 under the title De Jure Belli ac Pacis. *For the rest of his life he took a prominent part in public affairs and particularly in diplomatic missions. After an unsuccessful attempt at compromise by Grotius, Oldenbarneveldt was overthrown by Prince Maurice and the militant Calvinist faction in 1618 and Grotius was sentenced to life imprisonment. He managed to escape to France hidden in a chest and there continued his literary and scientific work with a French pension. He declined high appointments from the government of his own country in later years but in 1634 became Swedish ambassador to France. He used his position to try to bring about, without success, a negotiated end of the Thirty Years' War. He died at Rostock in 1645.*

The Civil Law, both that of Rome and that of each nation, in particular, has been treated of with a view to illustrate it or to present it in a compendious form by many. But International Law, that which regards the mutual relations of several Peoples, whether it proceeds from nature or be instituted by divine command or introduced by custom and tacit compact, has been touched on by few and has been by no one treated as a whole in an orderly manner. And yet that this be done concerns the human race.

I, for reasons which I have stated, holding it to be most certain that there is among nations a common law of Rights which is of force with regard to war, and in war, saw many and grave causes why I should write a work on that subject. For I saw prevailing throughout the Christian World a license in making war of which even barbarous nations would have been shamed; recourse being had to arms for slight reasons or no reasons; and when arms were once taken up, all reverence for divine and human law was thrown away, just as if men were henceforth authorized to commit all crimes without restraint.

In the first principle of nature (self-preservation) there is nothing which is repugnant to war; indeed all things rather favour it: for the end of war, the preservation of life and limb, and the retention or acquisition of things useful to life, argues entirely with the principle. . . .

Again, Right Reason and the nature of Society, which are next to be considered, do not prohibit all force, but that only which is repugnant to Society, that is, that which is used to attack the Rights of others. For Society has as its object that every one may have that which is his own in safety, by the common help and agreement.

The last and widest reason for taking up arms is the connection of men with men as such which alone is often sufficient to induce them to give their aid. Men are made for mutual help, says Seneca and the like. Here the question is raised whether man be bound to defend man, and people to defend people, from wrong. . . . There is another question, whether a war for the subjects of another be just, for the purpose of defending them from injuries inflicted by their ruler.

The Corinthians in Thucydides say that it is right that each state should punish its own subjects. . . . But all this applies when the subjects have really violated their duty; and we may add, when the case is doubtful. . . . But the case is different if the wrong be manifest.

Thus Seneca thinks that I may attack in war him who, though a stranger to my nation, persecutes his own; as we said when we spoke of exacting punishment: and this is often joined with the defence of innocent subjects. We know indeed, both from ancient and from modern histories, that the desire to appropriate another's possessions often uses such a pretext as this but that which is used by bad men does not necessarily therefore cease to be right. Pirates use navigation, but navigation is not therefore unlawful. Robbers use weapons, but weapons are not therefore unlawful.

But, as we have said, that leagues made with a view to mutual help in all wars alike, without distinction as to the cause, are unlawful; so no kind of life is more disreputable than that of those who act as soldiers for pay merely, without regard to the cause; whose motto is, the right is where the best pay is. . . .

* * *

War is not one of the acts of life. On the contrary, it is a thing so horrible that nothing but the highest necessity or the deepest charity can make it right. . . .

As Sallust says, a good man takes up the beginning of war reluctantly and does not follow its extremes willingly. This of itself ought to be enough; but often human Utility draws men the same way; those first who are weakest; for a long struggle with a powerful adversary is perilous, and as in a ship we must avert a greater calamity by some less, putting away anger and hope, fallacious advisers, as Livy says. So Aristotle. But also this is for the benefit of the stronger: for as Livy also says: To them peace if they grant it is bounteous and creditable and better than a victory merely hoped for. For they must recollect that Mars is on both sides. So Aristotle. And so in the oration of Diodorus. And there is much to be feared from the courage of despair, like the dying bites of a wild animal.

If the two reckon themselves equal, then, as Caesar holds, is the best time for treating of peace, since each trusts in himself. Peace made on any condition whatever is, by all means, to be kept, on account of the sacredness of good faith, of which we speak; and care must be had to avoid not only perfidy but anything which may exasperate the mind of the other party.

* * *

But if they are commanded to join in a war, as often happens, if they are quite clear that the war is unlawful, they ought to abstain. That God is to be obeyed rather than men, not only the Apostles have said but Socrates also.

The rule is the same if anyone be falsely persuaded that what is commanded is unjust. The thing is unlawful for him, as long as he re-tains the opinion, as appears by what is said above. But if the subject doubts whether the matter be lawful or not, must he remain quiet or obey (and assist in war)? Most writers think that he ought to obey. And they hold that the rule does not apply—if you doubt, do not do it. Because he who doubts speculatively may not be in doubt in his practical judgment. He may believe that in a doubtful matter he ought to obey his superior. . . . They say, he does the damage who orders it to be done; he is no fault who is obliged to obey; the necessity imposed by authority excuses; and the like.

But if the minds of the subjects cannot be satisfied by the expositions of the cause, it will by all means be the part of a good magistrate rather to impose extraordinary contributions upon them, than military service, especially as it is to be supposed that persons willing to serve as soldiers will not be wanting, whose acts not only if they are morally good but even if they are bad a just king may use, even as God makes use of the spontaneous acts of the devil and of impious men; and he is free from fault who, being in pecuniary distress, takes money from a wicked usurer.

And even if there can be no doubt as to the justice of the war, it does not seem at all equitable that Christians who are unwilling should be compelled to act as soldiers.

I think however, that it may happen that in a war not doubtful but even manifestly unjust there may be a just defense of the subjects who take a part in it. . . . It follows that if it be clear that the enemy comes with such a purpose that though he could save the lives of the subjects of his adversary, he will not, those subjects may defend themselves by the law of nature which they are not divested of by the law of nations.

THOMAS HOBBES:

FROM THE

Leviathan

Thomas Hobbes was born at Malmesbury in 1588 and after graduating from Oxford became a private tutor. He also worked for a time for Sir Francis Bacon and translated Thucydides. His views were so unpopular that he fled to France in 1640, where he tutored the exiled Charles II, returning to England in 1651. In 1662, after the Restoration, Parliament ordered an investigation of his atheistic works and Hobbes agreed not to publish anything further of a provocative nature. He continued to write prolifically, however, until the end of his long life in 1679.

Hereby it is manifest that during the time men live without a common Power to keep them all in awe, they are in that condition which is called War, and such a war as is of every man, against everyman. For War consisteth not in Battle only, of the act of fighting; but in a tract of time, wherein only the Will to contend by battle is sufficiently known: and therefore the notion of Time is to be considered in the nature of War, as it is in the nature of Weather. For as the nature of Foul Weather lieth not in a shower or two of rain, but in an inclination thereto of many days together, so the nature of war consisteth not in actual fighting, but in the known disposition thereto during all the time there is no assurance to the contrary. All other time is Peace.

Whatsoever, therefore, is consequent to a time of War, where everyman is Enemy to every man, the same is consequent to the time wherein men live without other security than what their own strength and their own invention shall furnish them withal. In such condition there is no place for Industry, because the fruit thereof is uncertain: and consequently no Culture of the Earth; no Navigation, nor use of the commodities that may be imported by the Sea, no commodious Building, no Instruments of moving—and removing—such things as require much force; no Knowledge of the face of the Earth; no account of Time, no Arts, no Letters; no Society; and which is worst of all, continual fear and danger of violent death; and the life of man, solitary, poor, nasty, brutish, and short.

OLIVER CROMWELL:

FROM

Letter to the Honourable William Lenthall Esq.

Oliver Cromwell was born in Huntingdon in 1599 and engaged in farming after attending Cambridge. He became a Puritan and a member of Parliament. At the outbreak of the Civil War he became a soldier at the age of forty-three without previous military experience, and henceforth his career was one of unbroken military triumph: Marston Moor, Naseby, Preston, Dunbar, and Worcester. The Royalists were crushed, the Scots routed, and the Irish subjugated. As commander in chief of the New Model Army, Cromwell dissolved Parliament in 1653 and was proclaimed Lord Protector. His forces defeated the Dutch and Spanish, intimidated the French, and controlled the country at home until Cromwell died, a disillusioned man, in 1658.

After the resumption of the Civil War, the Scots were crushed at Preston in 1648 and then, after an evasive campaign by Cromwell, finally confronted on the east coast of Scotland. What follows is an excerpt from Cromwell's letter to the Speaker of the Parliament of England.

At a general council, it was thought fit to march to Dunbar, and there to fortify the town, which we thought, if any thing would provoke them to engage, as also that the having of a garrison there, would furnish us with accommodation for our sick men, and would be a good magazine (which we exceedingly wanted) being put to depend upon the uncertainty of weather for landing provisions, which many times cannot be done, though the being of the whole army lay upon it (all the coasts from Berwick to Leath having not one good harbour).

As also to lie more conveniently to receive our recruits of horse and foot from Berwick.

Having these considerations, upon Saturday, the 30. of August, we marched from Musselborough to Haddington, where by the time we had got the van brigade of our horse, and our foot, and train into their quarters: the enemy was marched with that exceeding expedition, that they fell upon the rear forlorn of our horse, and put it in some disorder, and indeed, had like to have engaged our rear brigade of horse, with their whole horse, had not the Lord by his providence put a cloud over the moon, thereby giving us opportunity to draw off those horse to the rest of the army.

Which accordingly was done without any loss, save 3 or 4 of our own aforementioned forlorn, wherein the enemy (as we believe) received more loss.

The army being put into a reasonable secure posture; towards midnight the enemy attempted our quarters, on the west end of Haddington, but through the goodness of God we repulsed them.

The next morning we drew into an open field, on the south side of Haddington, we not judging it safe for us to draw to enemy upon his own ground, he being prepossessed thereof, but rather drew back to give him way to come to us, if he had so thought of it.

And having waited about the space of 4 or 5 hours, to see if he would come to us; and not finding any inclination in the enemy so to do, we resolved to go according to our first intendment to Dunbar. . . .

The enemy that night we perceived gathered towards the hills, labouring to make a perfect interposition between us and Berwick, and having in this posture a great advantage, through his better knowledge of the country which he effected by sending a considerable party to the strait passage at Copperspath, where ten men to hinder, are better than forty to make their way. . . .

Upon Monday evening the enemies whole numbers were very great (as heard about 6000 horse and 16000 foot at least), ours drawn down, as to sound men, to about 7500 foot, and 3500 horse.

The enemy drew down to the right wing about two-thirds of their left wing of horse, to the right wing, shogging also their foot and train much to the right; causing their right wing of horse to edge down towards the sea.

We could not well imagine, but that the enemy intended to attempt upon us, or to place themselves in a more exact condition of interposition; the Major General and myself coming to the Earl of Roxborough's house, and observing this posture, I told him, I thought it did give us an opportunity and advantage to attempt upon the enemy, to which he immediately replied; that he had thought to have said the same thing to me.

So that it pleased the Lord to set this apprehension upon both of our hearts, at the same instant. We called for Colonel Monke and shewed him the thing and coming to our quarters at night, and demonstrating our apprehensions to some of the Col. they also cheerfully concurred.

We resolved therefore to put our business into this posture, that six regiments of horse, and three regiments and a half of foot should march in the van: and that the Maj. Gen. the Lieut. Gen. of the horse, and Commis. General and Colonel Monke to command the brigade of foot, should lead on the business: and that Colonel Pride's brigade, Colonel Overton's brigade, and the remaining two regiments of horse should bring up the cannon and rear; the time falling on to be by break of day, but through some delays it proved not to be so, till six o'clock in the morning.

The enemy's word was "the Covenant" which it had been for divers days. Ours, "the Lord of Hosts."

* * *

The horse in the mean time did with a great deal of courage and spirit, beat back all oppositions, charging through the bodies of the enemies horse, and foot, who were after the first repulse given, made by the Lord of Hosts, as stubble to their swords.

* * *

Thus you have the prospect of one of the most signal mercies God hath done for England and his people this war; and now it may please you to give me leave of a few words.

It is easy to say the Lord hath done this; it would do you good to see and hear our poor foot to go up and down making their boast to God. But, Sir, it is in your hands, and by these eminent mercies of God puts it more into your hands, to give glory to him to improve your power and his blessings to his praise. We that serve you beg of you not to own us, but God alone; we pray you own his people more and more, for they are the chariots and horse men of Israel. Disown yourselves, but own your authority, and improve it to curb the proud and the insolent, such as would disturb the tranquillity of England, though under what specious pretences soever.

Relieve the oppressed, hear the groans of poor prisoners in England, be pleased to reform the abuses of all professions; and if there be anyone that makes many men poor to make a few rich, that suits not a commonwealth.

Beseeching you to pardon this length, I humbly take leave and rest, Sir,

Your most humble servant,
O. Cromwell
Dunbar, 4th of September, 1650

MARQUIS OF HALIFAX:

FROM

Essays, "Rough Draft for a New Model at Sea"

"I will make no introduction to the following discourse than that as the importance of our being strong at sea was ever very great . . . so now it is become indispensably necessary for our very being."

George Savile was born in Yorkshire in 1633 and was brought up partly on the Continent. He was elected to the Parliament of 1660 which restored Charles II. During the ensuing reigns he shifted his political position in a way which earned him the soubriquet of "the Trimmer." In 1682 he was appointed Lord Privy Seal and created Marquis of Halifax, but was dismissed by James II in 1685. In 1689 he was instrumental in the acceptance of William and Mary as joint sovereigns and was appointed chief minister. He resigned in 1690 and died in 1700. During his periods of retirement he wrote his political essays, embodying his philosophy of constitutional and religious compromise.

The discourse about the roles of "gentlemen" and "sailors" had gone on in Elizabethan and sixteenth century England as it had, with Plato and Aristotle, in ancient Greece.

It appeareth then that a bounded monarchy is that kind of government which will most probably prevail and continue in England from whence it must follow that every considerable part ought to be composed, as the better to conduce to the preserving the harmony of the whole constitution. The Navy is of so great importance that it would be disparaged by calling it less than the life and soul of government.

Therefore to apply the argument to the subject we are upon; in case the officers be all tarpaulins, it would be in reality too great a tendency to a commonwealth; such a part of the constitution being democratically disposed may be suspected to endeavour to bring it into that shape. In short, if the maritime force, which is the only thing that can defend us, should be wholly directed by the lower sort of men, with an entire exclusion of the nobility and gentry, it will not be easy to answer the arguments supported by so great a probability that such a scheme would not only lead toward a democracy, but directly lead us into it.

Let us now examine the contrary proposition, viz. that all officers should be gentlemen.

Here the objection lieth so fair, of its introducing an arbitrary government, that it is as little to be answered in that respect, as the former is in the other. Gentlemen, in a general definition, will be suspected to lie more than other men under the temptation of being made instruments of unlimited power.

The two former exclusive propositions being necessarily to be excluded in the question, there remaineth no other expedient than that there must be a mixture in the Navy of gentlemen and tarpaulins, as there is in the constitution of the government, of power and liberty. It is possible that the men of Wapping may think they are injured, by giving them any partners in the dominion of the sea. But I shall in a good measure reconcile myself to them by what follows; viz. the gentlemen shall not be capable of bearing office at sea, except they be tarpaulins too; that is to say, except they are so trained up by a continued habit of living at sea, that they may have a right to be admitted free denizens of Wapping.

When a gentleman is preferred at sea, the tarpaulin is very apt to impute it to friend or favour: but if that gentleman hath before his preferment passed through all the steps which lead to it, so that he smelleth as much of pitch and tar, as those that were swaddled in sail-cloth; his having an escutcheon will be so far from doing him harm, that it will set him upon the advantage ground: it will draw a real respect to his quality when so supported, and give him an influence and authority infinitely superior to that which the mere seaman can ever pretend to.

DANIEL DEFOE:

FROM

Memoirs of a Cavalier

*Daniel Defoe was born in London in 1659, the son of a
butcher, and educated at a Non-Conformist academy. He
joined Monmouth's rebellion in the West Country in 1683,
but managed to escape and was a volunteer trooper when
William of Orange entered London after the Glorious
Revolution. In 1698 he wrote an* Essay on Projects *with
a number of proposals for institutional reforms, including
military colleges, and an* Argument for a Standing Army.
*Thereafter he was a prolific writer of pamphlets and of
books.*

*In 1702 he was imprisoned, pilloried, and fined for a
religious satire and after his release in 1704 was employed
as a government agent in Scotland, working for Union. In
1715 he was again imprisoned for libel.* Robinson Crusoe
*was published in 1719. He was violently assaulted in 1724
in mysterious circumstances and died in debt in 1731.*

The Memoirs *were widely believed to be authentic (by
William Pitt, among others) but they were, in fact, one of
the earliest historical novels.*

Then I had the opportunity of seeing the Dutch Army, and their
famous general Prince Maurice. It is true that the men behaved themselves well enough in action, when they were put to it, but the
prince's way of beating his enemies without fighting, was so unlike the
gallantry of my royal instructor that it had no manner of relish with
me. Our way in Germany was always to seek out the enemy and fight

him, and, give the imperialists their due, they were seldom hard to be found, but were as free of their flesh as we were.

Whereas Prince Maurice would lie in a camp till he had starved half his men, if by lying there he could but starve two-thirds of his enemies; so that, indeed, the war in Holland had more of fatigue and hardships in it, and ours had more of fighting and blows. Hasty marches, long and unwholesome encampments, winter parties, counter-marching, dodging, and intrenching, were the exercises of his men, and often time killed more men with hunger, cold and diseases than he could do with fighting; not that it required less courage, but rather more, for a soldier had at any time rather die in the field a la coup de mousquet, than be starved with hunger, or frozen to death in the trenches.

Nor do I think to lessen the reputation of that great general, for it is most certain he ruined the Spaniards more by spinning the war thus out in length, than he could possibly have done by swift conquest; for had he, Gustavus like, with a torrent of victory, dislodged the Spaniard from all the twelve provinces in five years (whereas he was forty years in beating them out of seven) he had left them rich and strong at home, and able to keep the Dutch in constant apprehension of the return of his power; whereas, by the long continuance of the war, he so broke the very heart of the Spanish monarchy, so absolutely and irrecoverably impoverished them that they have ever since languished of the disease, till they are fallen from the most powerful to be the most despicable nation in the world.

III.

THE AGE
OF REASON

At the present day war is carried on by regular armies: the people, the peasantry, the townsfolk take no part in it and as a rule have nothing to fear from the sword of the enemy.

—Vattel, 1714–1767
Law of Nations

JONATHAN SWIFT:

FROM

The Conduct of the Allies

*Jonathan Swift was born in Dublin in 1667. From 1713
until his death he was Dean of St. Patrick's Cathedral, but
it was as a prose satirist that he established his enduring
reputation. Despite his Whig principles, he agreed in 1710
to become the chief political writer of the new Tory ministry
of Queen Anne, headed by Robert Harley and Henry St.
John.*

*Gulliver's Travels, Swift's best-known work, appeared
in 1726. He died in Dublin after long physical and mental
illness, in 1745.*

*Swift's pamphlet, commissioned by the new Tory Govern-
ment in 1711, was published on November 27. Thirty thousand copies were soon sold. Parliament reassembled on
December 7 and Marlborough was dismissed on December
30.*

The motives that may engage a wise prince or state in a war, I
take to be one or more of these: either to check the overgrown power
of some ambitious neighbour; to recover what has been unjustly
taken from them; to revenge some injury they have received, which
all political casuists allow; to assist some ally in a just quarrel, or,
lastly, to defend themselves when they are invaded. In all these cases,
the writers upon politics admit a war to be justly undertaken.

The last is, what has been usually called *pro aris et focis;* where
no expense or endeavour can be too great, because all we have is at
stake, and consequently our utmost force to be exerted; and the dis-

pute is soon determined, either in safety or in utter destruction. But in the other four, I believe, it will be found that no monarch or commonwealth did ever engage beyond a certain degree; never proceeding so far as to exhaust the strength and substance of their country by anticipations and loans, which, in a few years, must put them in a worse condition than any they could reasonably apprehend from those evils, for the preventing of which they first entered into the war; because this would be to run into real infallible ruin only in hope to remove what might, perhaps, appear so, by a probable speculation. . . .

Suppose the war to have commenced upon a just motive, the next thing to be considered is, when a prince ought in prudence to receive the overtures of a peace; which I take to be, either when the enemy is ready to yield the point originally contended for, or when that point is found impossible to be ever obtained; or when contending any longer, although with probability of gaining that point at last, would put such a prince and his people in a worse condition than the present loss of it. All which considerations are of much greater force where a war is managed by an alliance of many confederates, which, in a variety of interests among the several parties, is liable to so many unforseen accidents.

In a confederate war, it ought to be considered which party has the deepest share of the quarrel; for, although each may have their particular reasons, yet one or two among them will probably be more concerned than the rest, and therefore ought to bear the greatest part of the burden, in proportion to their strength. . . .

We have now for ten years together turned the whole force and expense of the war, where the enemy was best able to hold us at bay; where we could propose no manner of advantage to ourselves; where it was highly impolitic to enlarge our conquests; utterly neglecting that part which would have saved and gained us many millions; which the perpetual maxims of our government teach us to pursue; which would have soonest weakened the enemy and must either have promoted a speedy peace or enabled us to continue the war.

Those who are fond of continuing the war, cry up our constant success at a most prodigious rate, and reckon it infinitely greater than in all human probability we had reason to hope. Ten glorious campaigns are passed; and now at last, like a sick man, we are just expiring with all sorts of good symptoms.

I say not this by any means to detract from the army and its leaders. Getting into the enemy's lines, passing rivers and taking towns, may be actions attended with so many glorious circumstances; but when all this brings no real solid advantage to us; when it has no other end than to enlarge the territories of the Dutch and to increase the fame and wealth of our general; I conclude, however, this comes about, that things are not as they should be; and that surely our forces and money might be better employed, both towards reducing our enemy and working out some benefit to ourselves. But the case is still much harder; we are destroying many thousand lives, exhausting our substance, not for our own interest, which would be but common prudence, not for a thing indifferent, which would be sufficient folly, but perhaps to our own destruction, which is perfect madness. . . .

But great events often turn upon very small circumstances. It was the kingdom's misfortune that the sea was not the Duke of Marl-borough's element; otherwise the whole force of the war would infal-libly have been bestowed there, infinitely to the advantage of his country, which would then have gone hand in hand with his own. . . .

By agreement subsequent to the grand alliance we were to assist the Dutch with forty thousand men, all to be commanded by the Duke of Marlborough. So that whether this was prudently begun or not, it is plain that the true spring or motive of it was the aggrandiz-ing of a particular family; and in short, a war of the general and the ministry, and not of the prince or people; since those very persons were against it when they knew the power, and consequently the profit would be in other hands.

With these measures fell in all that set of people who are called the monied men; such as had raised vast sums by trading with stocks and funds, and lending upon great interest and premiums; whose per-petual harvest is war, and whose beneficial way of traffic must very much decline in peace.

In the whole chain of encroachments made upon us by the Dutch, which I have above deduced; and under those several gross impositions from other princes; if anyone should ask, why our gen-eral continued so easy to the last? I know of no other way so proba-ble or indeed so charitable to account for it, as by that unmeasurable love of wealth which his best friends allow to be his dominant pas-sion. However, I shall waive anything that is personal upon the sub-ject. I shall say nothing of those great presents made by several princes, which the soldiers used to call winter foraging, and said it was better than that of summer; and of two and a half per cent subtracted out of all the subsidies we pay on those parts, which amounts to no inconsiderable sum; and lastly, of the grand perquisites in a long and successful war, which are so amicably adjusted between him and the States.

The first overtures from France are made to England on safe and honourable terms; we who bore the brunt of the war ought in reason to have the greatest share in making the peace. If we do not hearken to a peace, others certainly will, and get the advantage of us there, as they have done in the war. We know that the Dutch have perpetually threatened us, that they will enter into separate measures of peace; and by the strength of that argument as well as by other powerful motives, prevailed on those who were then at the helm, to comply with them on any terms, rather than put an end to a war, which every year brought them such great accessions to their wealth and power.

MAURICE DE SAXE:

FROM

Reveries

Hermann Maurice Comte de Saxe was born in Germany in 1696, one of the 374 acknowledged bastards of the Elector Frederick August of Saxony, later King of Poland. He was commissioned at the age of twelve, fought under Prince Eugene at the battle of Malplaquet, and was given command of a German regiment, which had been bought for him, at the age of seventeen. He began to experiment with new methods of infantry training. In 1717 he took part in the capture of Belgrade from the Turks and in 1726 was elected Duke of Courland. He was expelled by the Russians the following year and returned to France.

He took part in the War of the Polish Succession, becoming a lieutenant-general in 1734 and in the War of the Austrian Succession, capturing Prague in 1741. In 1744 he planned an invasion of Britain in support of the Young Pretender, but a storm shattered the project. The same year he became a marshal of France and the following year he won the battle of Fontenoy against the British and their allies. He captured Brussels, and in 1747 was appointed marshal-general of France, with almost complete power over the Netherlands. After the successful conclusion of his campaigns and the Peace of Aix-la-Chapelle, he retired in ill health to the Chateau of Chambord, where he maintained a private regiment and a Negro bodyguard. He died there in 1750.

"Life is but a dream," he said to his doctor at the end, "but mine has been a fine one."

I have formed a picture of a general commanding which is not chimerical—I have seen such men.

The first of all qualities is COURAGE. Without this the others are of little value, since they cannot be used. The second is INTELLIGENCE, which must be strong and fertile in expedients. The third is HEALTH.

He should possess a talent for sudden and appropriate improvisation. He should be able to penetrate the minds of other men, while remaining impenetrable himself. He should be endowed with the capacity of being prepared for everything, with activity accompanied by judgment, with skill to make a proper decision on all occasions, and with exactness of discernment.

The functions of a general are infinite. He must know how to subsist his army and how to husband it; how to place it so that he will not be forced to fight except when he chooses; how to form his troops in an infinity of different dispositions; how to profit from that favorable moment which occurs in all battles and which decides their success. All these things are of immense importance and are as varied as the situations and dispositions which produce them.

In order to see all these things the general should be occupied with nothing else on the day of battle. The inspection of the terrain and the disposition of his troops should be prompt, like the flight of an eagle. This done, his orders should be short and simple, as for instance: "The first line will attack and the second will be in support."

The generals under his command must be incompetent indeed if they do not know how to execute this order and to perform the proper maneuvers with their respective divisions. Thus the commander in chief will not be forced to occupy himself with it nor be embarrassed with details. For if he attempts to be a battle sergeant and be everywhere himself, he will resemble the fly in the fable that thought he was driving the coach.

Thus, on the day of battle, I should want the general to do nothing. His observations will be better for it, his judgment will be more sane, and he will be in better state to profit from the situations in which the enemy finds himself during the engagement. And when he sees an occasion, he should unleash his energies, hasten to the critical point at top speed, seize the first troops available, advance them rapidly, and lead them in person. These are the strokes that decide battles and gain victories. The important thing is to see the opportunity and to know how to use it.

Prince Eugene possessed this quality, which is the greatest in the art of war and which is the test of the most elevated genius. I have applied myself to the study of this great man and on this point can venture to say that I understand him.

Many commanding generals only spend their time on the day of battle in making their troops march in a straight line, in seeing that they keep their proper distances, in answering questions which their aides de camp come to ask, in sending them hither and thither, and in running about incessantly themselves. In short, they try to do everything and, as a result, do nothing. They appear to me like men with their heads turned, who no longer see anything and who only are able to do what they have done all their lives, which is to conduct troops methodically under the orders of a commander. How does this happen? It is because very few men occupy themselves with the higher problems of war. They pass their lives drilling troops and believe that this is the only branch of the military act. When they arrive at the command of armies they are totally ignorant, and, in default of knowing what should be done, they do what they know.

One of the branches of the art of war, that is to say drill and the method of fighting, is methodical; the other is intellectual. For the conduct of the latter it is essential that ordinary men should not be chosen.

Unless a man is born with talent for war, he will never be other than a mediocre general. It is the same with all talents; in painting, or in music, or in poetry, talent must be inherent for excellence. All sublime arts are alike in this respect. That is why we see so few outstanding men in a science. Centuries pass without producing one. Application rectifies ideas but does not furnish a soul, for that is the work of nature.

I have seen very good colonels become very bad generals. I have known others who were great takers of villages, excellent for maneuvers within an army, but who, outside of that, were not even able to lead a thousand men in war, who lost their heads completely and were unable to make any decision.

If such a man arrives at the command of an army, he will seek to save himself by his dispositions, because he has no other resources. In attempting to make them understood better he will confuse the spirit of his whole army with multitudinous messages. Since the least circumstances changes everything in war, he will want to change his arrangements, will throw everything in horrible confusion, and infallibly will be defeated.

I do not favour pitched battles, especially at the beginning of a war, and I am convinced that a skillful general could make war all his life without being forced into one.

Nothing so reduces the enemy to absurdity as this method; nothing advances affairs better. Frequent small engagements will dissipate the enemy until he is forced to hide from you.

I do not mean to say by this that when an opportunity occurs to crush the enemy that he should not be attacked, nor that advantage should not be taken of his mistakes. But I do mean that war can be made without leaving anything to chance. And this is the highest point of perfection and skill in a general. But when a battle is joined under favourable circumstances, one should know how to profit from victory and, above all, should not be contented to have won the field of battle in accordance with the present commendable custom.

—THOMAS R. PHILLIPS
(translator)

JEAN DE BOURCET:

FROM

Principles of Mountain Warfare

Jean de Bourcet was born in 1700 and joined the French army in 1719 as an engineer officer. Like his famous predecessor, Vauban, Bourcet devoted most of his long military career to the principles and active conduct of siege warfare. At the same time he exercised a profound influence on general organization and strategical ideas of the French army—to which Napoleon was to be indebted. In 1742, at the start of the Piedmont campaign, he was chief engineer of the fortress of Mont Dauphin. He was responsible for the successful plans for the defense of southern France, culminating in the victory of Bassignano. He died in 1780.

In a mountain region, the all-important points for military purposes are the defiles, and when these, as is frequently the case, are impregnable against frontal attacks, the general taking the offensive must seek every possible means of turning them, and must so arrange his troops as to fix the enemy's attention on some point other than that of which it is intended to gain possession.

For this purpose, a general will do well to divide his army into a number of comparatively small bodies, a proceeding which in another kind of country would be dangerous, but which in the mountains is indispensable and safe provided the general who adopts it makes such arrangements that he can reunite his forces the moment that becomes necessary. He must therefore make his dispositions so that the enemy cannot interpose between the fractions into which his army is divided. . . .

A general who intends to take the offensive should assemble his army in three positions, distant not more than a march from one another, for in this way, while he will threaten all points accessible from any portion of the 25 or 30 miles thus held, he will be able suddenly to collect his whole army either in the centre or on either wing. The enemy will then be tempted to post troops to defend each of the threatened avenues of approach, and the attempt to be strong at all points will make him weak at each separate portion.

However carefully the enemy may have prepared his communications between several parts of his army, and have drafted the orders for its reunion in case of an attack at any point he will not be able to concentrate his troops there in time, if only the attacking general has concealed his plan and his first movements. The attacking general will usually be able to steal a march, if need be by moving at night, while the defender requires time to receive warning, time to issue his orders, and time for the march of the troops to the point attacked.

—SPENCER WILKINSON
(translator)

FREDERICK THE GREAT:

FROM

Memoirs;

FROM

Military Instructions

Frederick was born in Berlin in 1712, the heir to Frederick
William I of Prussia. After disagreements with his father
and an unsuccessful attempt to escape to England, he was
court-martialed and imprisoned. After gradual restoration
to favor, he was attached to the army of Prince Eugene
and developed a close friendship with Voltaire and other
leading intellectuals. Upon his accession in 1740 he seized
Silesia. After a series of brilliant victories he was left in
possession of Silesia by the Peace of Dresden in 1745. He
devoted the next ten years to the consolidation of his state,
including his army. After the outbreak of the Seven Years'
War in 1756 he found himself at overwhelming odds with
France, Russia, and Austria. Despite some initial victories,
only the unexpected accession of the admiring Peter III in
Russia saved him from total disaster and enabled him to
retain Silesia in 1763. Thereafter he concentrated principally on domestic reform and died, virtually alone, in his
palace of Sans Souci in 1786.

FROM *Memoirs*

What is the use of life if one merely vegetates? What is the point
of seeing if one only crams facts into his memory? In brief, what

good is experience if it is not directed by reflection? Vegetius stated that war must be a study and peace an exercise, and he is right.

Experience deserves to be investigated, for it is only after repeated examinations of what one has done that the artists succeed in understanding principles and in moments of leisure, in times of rest, that new material is prepared for experiment. Such investigations are the product of an applied mind, but this diligence is rare and, on the contrary, it is common to see men who have used all their limbs without once in their lives having utilized their minds. Thought, the faculty of combining ideas, is what distinguishes man from a beast of burden. A mule who has carried a pack for ten campaigns under Prince Eugene will be no better a tactician for it, and it must be confessed, to the disgrace of humanity, that many men grow old in an otherwise respectable profession without making any greater progress than this mule.

To follow the routine of the service, to become occupied with the care of its fodder and lodging, to march when the army marches, camp when it camps, fight when it fights—for the great majority of officers this is what is meant by having served, campaigned, grown gray in the harness. For this reason one sees so many soldiers occupied with trifling matters and rusted by gross ignorance. Instead of soaring audaciously among the clouds, such men know only how to crawl methodically in the mire. They are never perplexed and will never know the causes of their triumphs or defeats.

* * *

Scepticism is the mother of security. Even though only fools trust their enemies, prudent persons never do. The general is the principal sentinel of the army. He should always be careful of its preservation and that it is never exposed to misfortune. One falls into a feeling of security after battles, when one is drunk with success, and when one believes the enemy completely disheartened. One falls into a feeling of security when a skillful enemy amuses you with pretended peace proposals. One falls into a feeling of security by mental laziness and through lack of calculation concerning the intentions of the enemy.

To proceed properly it is necessary to put oneself in his place and say: what would I do if I were the enemy? What project could I form? Make as many as possible of these projects, examine them all, and above all reflect on the means to avert them. . . . But do not let these calculations make you timid. Circumspection is good only up to a certain point.

FROM *Military Instructions*

Here are the particulars for offensive warfare.

1. Your strategy must pursue an important objective. Undertake only what is possible and reject whatever is chimerical. If you are not fortunate enough to follow a great plan through to its perfection, you will nevertheless go much farther than the generals who, acting without plan, make war from day to day. Give battle only when you have reason to hope that your success will be decisive, and fight not only to defeat the enemy but to execute the course of your strategy that would fail but for this decision.

2. Never deceive yourself, but picture skillfully all the measures that the enemy will take to oppose your plans, in order never to be caught by surprise. Then, having foreseen everything in advance, you will already have remedies prepared for any eventuality.

3. Know the mind of the opposing generals in order better to divine their actions, to know how to force your actions upon them, and to know what traps to use against them.

4. The opening of your campaign must be an enigma for the enemy, preventing him from guessing the side on which your forces will move and the strategy you contemplate.

5. Always attempt the unexpected: this is the surest way to achieve success.

In a war between equals:

1. The more you employ stratagems and ruses, the more advantages you will enjoy over the enemy. You must deceive him and induce him to make mistakes in order to take advantage of his faults.

2. Always have as a goal to transform the war into an offensive on your part as soon as the occasion presents itself. All your maneuvers must lead toward this end.

3. Consider all the mischief that the enemy can do to you and prevent it by your prudence.

4. Do not attack the enemy when he adheres to the rules of war,

but profit from his slightest mistakes without delay. Whoever lets the occasion escape is not worthy of seizing it.

5. Profit from the battles you win, follow the enemy to the utmost, and push your advantages as far as you can extend them, because such happy events are not common.

6. Leave as little to fortune as possible by your foresight—chance will still have too much influence in military operations. It is enough that your prudence shares the stage with chance.

7. To win advantages over the enemy you must procure them, both by a war of partisans and by defeating his escorts, seizing his provisions, surprising his magazines, often defeating his detachments . . . [and] his rear guard, attacking him on the march, and finally, by engaging in battle with him, if he is badly posted, and even by surprising his winter quarters and falling on his posts if he has not provided for the security of his cantonments during the winter.

Here are the general maxims you must observe for defensive warfare.

1. Intend to put all your resources to work to change the nature of this war.

2. Anticipate everything detrimental the enemy can plan against you, and study expedients to elude his designs.

3. Choose impregnable camps that can contain the enemy by threatening his rear in the event that he changes posts, and be sure to cover your own magazines.

4. Accumulate many small advantages which, taken together, are the equivalent of great advantages. Try to make the enemy respect you in order to contain him by the fear of your arms.

5. Calculate all your movements carefully and observe the maxims and rules of tactics and castrametation strictly.

6. If you have advantages, make the most of them, and punish the enemy for his slightest errors, as though you were a pedagogue.

If you are on the defensive after losing a battle:

1. Your retreat must be short. You must get your troops accustomed again to looking the enemy in the face. Encourage them little by little, and wait for the proper moment to avenge your defeat.

2. Make use of ruses, stratagems, false information imparted to the enemy to lead to the happy moment when you can pay him back in his own coin—with interest—for the damage he did to you.

If you are less than half as strong as the enemy:

1. Wage partisan warfare: change the post whenever necessary.

2. Do not detach any unit from your troops because you will be beaten in detail. Act only with your entire army.

3. If you can throw your army against the enemy's communications without risking your own magazines, do so.

4. Activity and vigilance must be on the watch day and night at the door of your tent.

5. Give more thought to your rear than to your front, in order to avoid being enveloped.

6. Reflect incessantly on devising new ways and means of supporting yourself. Change your method to deceive the enemy. You will often be forced to wage a war of appearances.

7. Defeat and destroy the enemy in detail if it is at all possible, but do not commit yourself to a pitched battle, because your weakness will make you succumb. Win time—that is all that can be expected of the most skillful general.

8. Do not retreat toward places where you can be surrounded: remember [Charles XII at] Poltava without forgetting [the Duke of Cumberland at] Stade.

* * *

Of an army on the defensive awaiting reinforcements:

You risk everything by becoming involved in some undertaking before the juncture of your forces which, when united, would render you sure of whatever you would want to attempt. Thus you must confine yourself to the sphere of the strictest defensive during the period of concentration.

You can see by this presentation the extent to which the knowledge of a real general must be varied. He must have an accurate idea of politics in order to be informed of the intention of princes and the forces of states and of their communications; to know the number of troops that the princes and their allies can put into the field; and to judge the condition of their finances. Knowledge of the country where he must wage war serves as the base for all strategy. He must be able to imagine himself in the enemy's shoes in order to anticipate all the obstacles that are likely to be placed in his way. Above all, he must train his mind to furnish him with a multitude of expedients, ways and means in case of need. All this requires study and exercise. For those who are destined for the military profession, peace must be a time of meditation, and war the period where one puts his ideas into practice.

—JAY LUVAAS
(translator)

HENRY LLOYD:

FROM

The Military Rhapsody;

FROM

History of the Late War in Germany

"Land forces are nothing. Marines are the only species of troops proper for this nation; they alone can defend and protect it effectively."

Henry Lloyd (sometimes referred to as Henry Humphrey Evans) is believed to have been born in Merioneth in 1720, the son of a clergyman. The circumstances of his life are uncertain. He became a lay brother in a religious house in France after unsuccessfully trying to obtain a commission in the French army. Then he became military instructor to John Drummond of the Irish Brigade in the French army and obtained Marshal Saxe's permission to accompany the army as a mounted draftsman. As such he took part in the battle of Fontenoy.

He was appointed a third engineer and captain by the Young Pretender and took part in the 1745 Rising. He was wounded on board the Elizabeth *in its engagement with the* Lion *but reappeared in Carlisle with the Young Pretender. He went on a mission to Wales, trying to enlist support for the Jacobites and, secretly dressed as a priest, to travel to various western England ports—possibly with a view to a French landing. At the port of London he was arrested but released and returned to France. There he became a lieutenant colonel, and was sent back to England in 1754 to investigate further invasion possibilities.*

He served with the Prussian army, became a major general in the Austrian army and commanded a Russian division at the siege of Silistria. On at least one occasion he changed sides in the course of a war. He finally settled in Brussels where he died in 1783.

First published in 1779, the Rhapsody *appears to be derived from the author's experience with the invading forces during the Jacobite Rising of 1745—and thereafter as a secret agent. His point of view changed. And, apart from detailed defense considerations, he included some of his wider views of war.*

FROM *The Military Rhapsody*

The order of battle now adopted in Europe is, in many respects, defective and absurd. The infantry and cavalry, formed three deep, make the line so extensive that it loses all activity which is the soul of military manoeuvres and alone can insure success: insomuch that it may be established as an axiom that the army which moves and marches with the greatest velocity must, from that circumstance alone, finally prevail.

Our military institutions exclude every idea of celerity; hence it is that our victories are never complete and decisive, and that our attacks are reduced to some particular point which, gained or lost, the battle is over; the enemy retires, generally in good order, because from the extent and slowness of our motions we cannot pursue him with any vigour; he occupies some neighbouring hill and we have to begin again.

Moreover the position of the cavalry in a line on the flanks of the infantry is such that it retards the motions of the whole, because no one can advance unless the whole line does; besides, it cannot from that situation there support the infantry or be supported by it: the moment is lost before you can bring the cavalry where it is wanted. The reason assigned for placing the cavalry on the flanks is absurd viz. it covers the flanks of the infantry—Pray, is not the flank of the cavalry much weaker than that of the infantry? Since it cannot in any manner form a flank to protect itself, much less will it protect the flank of the infantry.

To remedy these defects I would humbly propose that all infantry be formed in such a manner that between each battalion or regiment, an interval of one hundred and fifty yards be left; behind these intervals I would have the cavalry placed in two lines at a proper distance, each squadron separately, with intervals to manoeuvre upon.

Whether you advance to the enemy or the enemy comes to you, the light troops disperse to the right and left, and you hear no more of them till the next day. Why on such occasions they do not form on the right and left of the army, at a convenient distance, and attack the enemy on the flanks, is to me inconceivable and the use now made of them appears ridiculous and absurd. Four or five hundred men; including one hundred hussars, distributed into small parties in the woods, behind the hedges, near the high roads would observe the enemy much better than ten thousand men. . . .

Land forces are nothing. Marines are the only species of troops proper for this nation; they alone can defend and protect it effectively.

FROM *History of the Late War in Germany*

Published in successive volumes in 1766 and 1782, this detailed study of campaigns in the Seven Years' War, in which the author had commanded on both sides, included the author's conclusions on eighteenth-century war.

It is universally agreed upon, that no art or science is more difficult than that of war; yet by an unaccountable contradiction of the human kind, those who embrace this profession take little or no pains to study it. They seem to think that the knowledge of a few insignificant trifles constitute a great officer. This opinion is so general that little or nothing else is taught at present in any army whatever. The continual changes and variety of motions, evolutions, etc. which soldiers are taught prove evidently they are founded on mere caprice.

This art, like all others, is founded on certain fixed principles, which are by their nature invariable, the application of them only can be varied: but they are in themselves constant.

This most difficult science may, I think, be divided into two; one mechanical, and may be taught by precepts; the other has no name, nor can it be defined or taught. It consists in a just application of the principles and precepts of war, in all the numberless circumstances and situations which may occur; no rule, no study, or application, however assiduous, no experience, however long, can teach this part; it is the effect of genius alone.

As to the first, it may be reduced to mathematical principles. Its object is to prepare the materials which form an army for all the different operations which may occur: genius must apply them according to the ground, number, species and quality of the troops, which admit of infinite combinations.

In this art as in poetry and eloquence, there are many who can trace the rules by which a poem or an oration should be composed, and even compose, according to the exactest rules; but for want of that enthusiastic and divine fire, their productions are languid and insipid: so in our profession, many are to be found who know every precept of it by heart; but alas! when called upon to apply them, are immediately at a stand. They then recall their rules and want to make everything, the rivers, the woods, ravines, mountains, etc. subservient to them; whereas their precepts should, on the contrary, be subject to these, who are the only rules, the only guide we ought to follow; whatever manoeuvre is not formed on them is absurd and ridiculous.

These form the great book of war; and he who cannot read it, must be forever content with the title of a brave soldier and never aspire to that of a great general.

* * *

The king's conduct was founded on the most sublime principles of war. Though his army was much inferior to that of the enemy, yet, by dint of superior manoeuvres, he brought more men into action, at the point attacked, than they; which must be decisive when the troops are nearly equal in goodness.

Wherefore generals must make it their study to establish in time of peace, such evolutions as facilitate the manoeuvre of armies, and, in time of war, choose such a field of battle, if possible, as enables them to hide part of their motions, and so bring more men into action than the enemy; and if the ground, either by its nature or by the vigilance of the enemy, does not permit them to cover their motions, then a greater facility of manoeuvring will answer the same end and enable them to bring more men to the PRINCIPAL POINT ATTACKED THAN THE ENEMY. The only advantage of a superior army in a day of action consists in this only, that the general can bring more men into action than the enemy; but if they do not move with facility and quickness and are not all brought to action at the same time, that superiority of numbers will serve only to increase the confusion.

From when we will deduce a general rule—"That general, who, by the facility of his motions, or by his artifice, can bring most men into action, at the same time, and at the same point, must, if the troops are equally good, necessarily prevail and therefore, all evolutions, which do not lend to this object, must be exploded."

* * *

People talk very much of the shock of cavalry. If they mean that the horses push each other and strike with their breasts, which the French who abound in unmeaning words, call Coup de Pontrail, it is an absurdity. . . .

Indeed, our battles, as we have seen, are commonly nothing more than great skirmishes, and therefore, as I have said before, wars are not now as formerly, concluded by battles, but for want of means to protract them.

* * *

Such [the American War of Independence] has been, and generally must be, the issue of wars prosecuted at a great distance, unless the first campaign gives you a decisive superiority: it follows, of course, that the success of such enterprises depends entirely on the vigour of your operations: if in the beginning they are not decisive, they never will be so hereafter.

* * *

Velocity is everything in war, particularly if the country be open and fruitful like Poland. Such an army, as we suppose, with two hundred carpenters, and ropes to make rafts, would ruin any European Army in a month. The Tartars have overcome and conquered a great part of the world by their velocity alone, whereas European Armies have not in two centuries conquered any one province of considerable extent, because they are too heavy.

* * *

There is nothing performed by contractors which may not be much better executed by intelligent officers. They make immense fortunes at the expense of the state which ought to be saved. They destroy the army, horse and foot and even the hospitals, by furnishing the worst of everything.

* * *

The French are gay, light and lively, governed rather by immediate and transitory impulse than by any principle or sentiment: their sensations, from the nature of their climate are very delicate, and therefore objects make a very strong impression, but momentary, because a new object, producing a new impression, effaces the former: from whence follows they are impetuous and dangerous in their attacks, all the animal spirits seem united, and produce a sort of furious convulsion, and gives them a more than ordinary degree of vigour for that instant, but it exhausts the whole frame: the instant following they appear languid and weak, and changed into other men.

Wherefore it should be a maxim, in making war against the French, to keep them continually in motion, especially in bad weather, always attack them, never permit them to follow their own dispositions, force them to observe yours; their impatience will soon reduce them to commit some capital error. If their leader is wise and prudent, and refuses to comply with their unreasonable requests, they will treat him with contempt, grow turbulent, and desert. The present ministry endeavours to introduce German discipline among them, without considering the difference there is between national characters, and I doubt whether it will produce the effects they expect from it. Nature must be improved, not annihilated.

* * *

But experience has proved that the Russian infantry is by far superior to any in Europe, insomuch that I question whether it can be defeated by any infantry whatsoever, and as their cavalry is not so good as that of other nations, reason dictates that a mixed order of battle alone can conquer them. They cannot be defeated, they must be killed, and infantry, mixed with great corps of cavalry, only can do this.

They made war, and always will, in all probability, like the Tartars. They will over-run a country, ravage and destroy it, and so leave it; because they can never, according to the method they now follow, make a solid and lasting conquest. They put themselves an insurmountable barrier to it. Their own light troops, and the want of a solid plan of operations, will one day ruin their army.

JAMES WOLFE:

FROM

Letter to Major Rickson, 5th November 1757, after the Expedition to La Rochelle

James Wolfe was born in Kent in 1726. He was commissioned in the marines in 1741, but transferred to an infantry regiment and took part in the War of the Austrian Succession, fighting against the French at Dettingen in 1743. He then took part in the suppression of the Jacobite uprising, fighting at Culloden where he was commended by the Duke of Cumberland. He was promoted to command a regiment in 1750 at the age of twenty-four. After the outbreak of the Seven Years' War he was sent to North America and captured Louisburg in a daring amphibious operation. On his return to England in 1758, he was promoted to major general and put in command of the expedition to capture Quebec. The following year, after an initial setback and while seriously ill with tuberculosis, he took the city by surprise, storming the Heights of Abraham and insuring British supremacy in North America. He was killed in the assault.

During his short career Wolfe had undertaken intensive military studies and acquired a reputation for oddity. "Mad, is he?" the King is said to have commented. "I wish he would bite some of my other generals."

I thank you heartily for your welcome back. I am not sorry that I went, notwithstanding what has happened; one may always pick up something useful from amongst the most fatal errors. I have found out that an Admiral should endeavour to run into an enemy's port immediately after he appears before it; that he should anchor the transport ships and frigates as close as he can to the land; that he should reconnoitre and observe it as quick as possible, and lose no time in getting the troops on shore; that previous directions should be given in respect to landing the troops, and a proper disposition made for the boats of all sorts, appointing leaders and fit persons for conducting the different divisions.

On the other hand, experience shows me that, in an affair depending upon vigour and despatch, the Generals should settle their plan of operations, so that no time may be lost in idle debate and consultations when the sword should be drawn; that pushing on smartly is the road to success, and more particularly so in an affair of this nature; that nothing is to be reckoned an obstacle to your undertaking which is not found really so upon trial; that in war something must be allowed to chance and fortune, seeing it is in its nature hazardous, and an option of difficulties; that the greatness of an object should come under consideration, opposed to the impediments that lie in the way; that the honour of one's country is to have some weight; and that, in particular circumstances and times, the loss of a thousand men is rather an advantage to a nation than otherwise, seeing that gallant attempts raise its reputation and make it respectable; whereas the contrary appearances sink the credit of a country, ruin the troops, and create infinite uneasiness and discontent at home.

I know not what to say, my dear Rickson, or how to account for our proceedings, unless I own to you that there never was people collected together so unfit for the business they were sent upon—dilatory, ignorant, irresolute, and some grains of a very unmanly quality, and very unsoldier-like or unsailor-like. I have already been too imprudent; I have said too much, and people make me say ten times more than I ever uttered; therefore, repeat nothing out of my letter, nor name my name as author to any one thing.

The whole affair turned upon the impracticability of escalading Rochefort; and the two evidences brought to prove that the ditch was wet (in opposition to the assertions of the chief engineer, who had been in the place) are persons to whom, in my mind, very little credit should be given, without these evidences we should have landed, and must have marched to Rochefort, and it is my opinion that the place would have surrendered, or have been taken, in forty-eight hours. It is certain that there was nothing in all that country to oppose 9000 good foot—a million Protestants, upon whom it is necessary to keep a strict eye, so that the garrison could not venture to assemble against us, and no troops, except the militia, within any moderate distance of these parts.

Little practice in war, ease and convenience at home, great incomes, and no wants, with no ambition to stir to action are not the instruments to work a successful war withal; I see no prospect of better deeds. . . .

ALEXANDER SUVOROV:

FROM

The Science of Conquering

Alexander Suvorov was born in Moscow in 1730, of Swedish extraction. Although a frail child he made a hero of Charles XII of Sweden and studied the military classics. One of his mentors was General Hannibal, a Negro slave who had become a general in Peter the Great's army and lived to be ninety-two. Suvorov joined the army as a private soldier when he was seventeen and, despite his connections, chose to remain in the ranks until he was twenty-five. Meanwhile he published some poems and articles and was already considered something of an oddity. By 1757 he was a lieutenant colonel and two years later distinguished himself in the Seven Years' War. He became a student of partisan warfare and a protagonist of the offensive. He played an outstanding part in the Polish and Turkish wars, which extended the territory of Russia.

In old age he was sent as commander in chief to fight the armies of the French Revolution and conducted a skillful campaign in Italy. On his return he found himself in disgrace with the mad Tsar Paul and died soon afterward in 1800.

Suvorov had a lasting influence on the Russian army. "Long before Napoleon," writes Marshal Sokolovsky in a Soviet assessment, "Suvorov successfully massed troops instead of the linear tactics and cordon strategy dominant in Western European armies." The Science of Conquering was not, however, meant as a staff treatise, but as a collection of military precepts which could be memorized and acted on by peasant armies.

Fire seldom but accurately. Thrust the bayonet with force. The bullet misses, the bayonet doesn't. The bullet's an idiot, the bayonet's a fine chap. Stab once and throw the Turk off the bayonet. Bayonet another, bayonet a third; a real warrior will bayonet half a dozen and more. Keep a bullet in the barrel. If three should run at you, bayonet the first, shoot the second, and lay out the third with your bayonet. This isn't common but you haven't time to reload. . . .

In an open battle there are three attacks. Firstly, in the weaker flank. It's unwise to tackle a strong flank covered by a wood. A soldier can pick his way across a marsh but it's harder to cross a river— you can't run across without a bridge. This is useless unless the cavalry have room to use their sabres—otherwise they themselves are crushed together. Lastly, an attack in the rear, which is very good but only for a small corps. It is difficult for an army to wheel round behind the enemy without his noticing.

In a field battle, fight in line against regulars and in a square against the Turks—no columns. But even against the Turks it will happen that a square of 500 men will have to tear through a mass of five or seven thousand with the help of flanking squares. In this case it should form into columns—but up to now there has been no need for them. Then there are the godless, flighty, madcap French. They wage war on the Germans and others in columns. If we shall happen against them we shall have to hit them with columns as well.

* * *

[Storming Forts] Break through the abatis, throw hurdles over the wolf traps, run fast, jump over the palisade, throw down your fascines, go down into the ditch, place the ladders. Marksmen, cover the column; Fire at the heads appearing on the battlements. Columns, jump across the wall on to the parapet. Stick to the parapet; Form a line; put a guard on the powder magazines. Open the gates for the cavalry; The enemy runs into the town—turn his guns against him; Fire hard down the streets. Give them a lively bombardment. Don't go after them until ordered to go down into the town. Then cut up the enemy on the streets. Cavalry, cut them down; Don't go into the houses. Hit them in the open: storm the places where they have gathered, occupy the square and mount a guard. Put pickets by the gates, the ammunition stores and magazines immediately. . . .

There is an enemy greater than the hospital: the damned fellow who "doesn't know." The hint-dropper, the riddle-poser, the deceiver, the word-spinner, the prayer-skimper, the two-faced, the mannered, the incoherent. The fellow who "doesn't know" has caused a deal of harm . . . one is ashamed to talk about him. Arrest for the officer who "doesn't know" and house arrest for the field or general officer.

Training is light and lack of training is darkness. The problem fears the expert; If a peasant doesn't know how to plough, he can't grow bread. A trained man is worth three untrained: that's too little—say six, six is too little—say ten to one. . . . We will beat them all, roll them up, take them prisoner.

EDWARD GIBBON:

FROM

Autobiography

*Edward Gibbon was born near London in 1737 and was
sent to Switzerland after a temporary conversion to Ca-
tholicism at Oxford. He became a temporary captain in the
South Hampshire militia during the Seven Years' War and
in 1774 entered Parliament as a supporter of Lord North.
Meanwhile he had begun writing* The Decline and Fall of
the Roman Empire, *completed in 1787. He died in 1794.*

*During the Seven Years' War, with fears of an invasion,
Gibbon was called to militia service.*

The loss of so many busy and idle hours was not compensated
by any elegant pleasures and my temper was insensibly soured by the
society of our rustic officers who were alike deficient in the knowledge of scholars and the manners of gentlemen.

In every state there exists, however, a balance of good and evil.
The habits of a scholarly life were usefully broken by the duties of an
active profession: in the healthful exercise of the field I hunted with a
battalion instead of a pack, and that time I was ready, at any hour of
the day or night, to fly from quarters to London, from London to
quarters, on the slightest call of private or regimental business. But
my principal obligation to the militia was the making me an Englishman and a soldier. After my foreign education, with my reserved
temper I should long have continued a stranger to my native country,
had I not been shaken in this various scene of new faces and new
friends; had not experience forced me to feel the character of our

leading men, the state of parties, the forms of office, and the operation of our civil and military system.

In this peaceful service I imbibed the rudiments of the language and science of tactics, which opened a new field of observation and study. I diligently read and meditated the *Mémoires Militaires* of Quintis Icilius (Mr. Guichardt), the only writer who has united the merits of a professor and a veteran. The discipline and evolution of a modern battalion gave me a clearer notion of the Phalanx and the Legions, and the Captain of the Hampshire grenadiers (the reader may smile) has not been useless to the historian of the Roman Empire.

JACQUES ANTOINE HIPPOLYTE DE GUIBERT:

FROM

General Essay on Tactics

Jacques Antoine Hippolyte Comte de Guibert was born in 1743 and joined the French army as a young man. His Essai Général de Tactique, *published in 1770, made him a celebrity in Paris society. Appointed colonel in command of the Corsican legion, he retired early from the army, engaged in pre-Revolutionary politics, and published another military work,* Defense du Système de Guerre Moderne *in 1779, which was to some extent a repudiation of the ideas in his earlier work. He also wrote tragedies in verse. Frustrated in his ambitions and in his personal life, he died in 1790 exclaiming "I shall be known! I shall receive justice!" After his death his widow sought to rescue him from oblivion by publishing his love letters to and from Mademoiselle de Lespinasse, which became a classic of their kind.*

First published and acclaimed in 1772, the General Essay on Tactics *was largely forgotten, like its author, by the Revolution. However, a new edition was authorized to state that "Bonaparte has carried the* Essai Général de Tactique *with him in the camps and has said that it is a book fit to form generals." It is considered to have been influential in forming Napoleon—the leader whom it predicted would arise.*

Tactics must be divided into two parts: one, elementary and limited, the other, compound and sublime.

The first comprises all the details of the formation, instruction and exercise of a battalion, a squadron, a regiment. So many sovereign ordinances, so many subordinate systems, so many contrary opinions exist in respect of it. It is what at present exercises our minds, and will long continue to do so, because the details are within everyone's grasp; because our national inconstancy, when not kept under control, varies its principles as well as its methods; and, finally, because to innovate or to associate with innovators has become a way to fame and fortune.

The second part is, strictly speaking, the science of Generals. It encompasses all the great features of war, such as the deployment of armies, orders of march, orders of battle. Thus, it shares in—and identifies itself with—the science of choosing positions and knowing the terrain, since this has no other purpose than to determine troop dispositions more surely. It is bound up with the science of fortification, since works should be built for troops and on a relative scale to them; it is bound up with artillery, since the deployment and execution of the latter should be combined with troop dispositions and deployment and since the artillery is only an accessory for the purpose of seconding and sustaining the troops.

Tactics, divided into two parts and developed as I conceive, are single and sublime. They become the science of all times, of all places, and of all arms; that is to say, if ever, by some unforeseeable revolution, in the type of our arms, we wanted to return to the "deep order," we could do so without changing either the deployment or the constitution of our forces. Tactics, in a word, are the result of all that the military centuries, before ours, have thought good and of what our century has been able to add. . . .

Let us suppose, on the side, an army overloaded with equipment, awkward to manoeuvre, as ours is, and, on the other side, a well-constituted army, mobile, commanded by a general who has pondered on all tactical possibilities. The first will seek positions and put all its trust in them; it will, slowly and with difficulty; it will be bound by the ways in which it ensures its provisioning, thinking itself lost if it has not drawn its train into exact position behind it. The second army will be light and mobile, capable of bold forced marches. It will always be on the offensive and will never close itself up in positions and will spurn those positions that others would like to impose on it. Will the enemy think they can stop it by one of these so-called inexpugnable positions? It will know how to conceal a move or, even without concealment, to come within sight of the enemy, on his flank or rear. To execute this movement, it will carry, if necessary, provisions for eight days and do without its train.

What will the enemy, astounded by this new kind of warfare, try to do? Will he wait for an army able to move swiftly, to switch in a moment from marching order to battle order? Will he wait for such an army to be in a position to attack the flank or rear of its opponent? Such inaction would prove fatal to him. Will he change his position? If so, he will lose the advantages of the ground on which he had relied and will be forced to accept battle where he can. . . .

In the end, I maintain that a well-constituted and well-commanded army should never find itself in front of a position which stops it or which compels it to attack at a disadvantage the army based there; except that it is one of those rare positions, adjacent to the site they aim to cover, which does not expose the possibility of a flank or rear manoeuvre. Such was the position so skillfully chosen by Marshal de Broglie in front of Frankfurt, and so gloriously justified by the victory which took place there. Such are also, generally speaking, the positions which an army can take up at the head of some unique pass which it wants to defend or in front of, or very close to, a place which the enemy will be forced to besiege. In any other case, I say, positions are to be spurned; it is easy to force the enemy to come out of them or, if they obstinately stay there, to attack them with advantage. One has only to bear down on his flank or rear, to attack him on any side other than the front, which is where he had planned his defensive disposition and where the ground is to his advantage.

I say that a general who, in this respect, throws off established prejudices, will embarrass and stun the enemy and give him no chance to breathe, forcing him to fight or to retreat continuously. But such a general would need an army differently constituted from our armies today, an army which, formed by himself, was prepared for the entirely new kind of operations which he would require it to perform.

JOHN PAUL JONES:

FROM

Letter to Vice-Admiral Kersaint, 1791

John Paul Jones was born in Scotland in 1747, the son of a gardener. At twelve he went to Virginia as a cabin boy and engaged for several years in the slave trade. He was arrested on a murder charge on two occasions but managed to escape, and in 1775 was commissioned in the new Continental navy, being the first to raise the Grand Union flag at sea. His daring exploits in the Revolutionary War, culminating in the engagement between his ship the Bonhomme Richard *and the* Serapis, *played an influential part in the recognition of American independence—and created a national legend.*

Shunned by Congress, he was received as a hero by French society. He was appointed an admiral by Catherine the Great to command a Russian fleet against the Turks but returned to France, where he died in 1792.

The letter was written not long before the "Father of the American Navy" died in poverty and neglect in revolutionary Paris.

I have noticed—and no reader of the naval history of France can have failed to notice it—that the underlying principle of operation and rule of action in the French Navy have always been calculated to subordinate immediate or instant opportunities to ulterior if not distant objects. In general I may say that it has been the policy of French admirals in the past to neutralise the power of their adversaries, if possible, by grand manoeuvres rather than to destroy it by grand attacks.

A case in point of this kind is the campaign of the Count de Grasse in his conjoint operation with the land forces of General Washington and the Count de Rochambeau, which so happily resulted in the capitulation of Cornwallis at Yorktown. . . .

Now, my dear Kersaint, you know me too well to accuse me of self-vaunting. You will not consider me vain, in view of your knowledge of what happened in the past off Carrickfergus, off Old Flaboro Head, and off the liman in the Black Sea, if I say that, had I stood—fortunately or unfortunately—in the shoes of de Grasse, there would have been disaster to some one off the Capes of the Chesapeake; disaster of more lasting significance than an orderly retreat of a beaten fleet to a safe port. To put it a little more strongly, there was a moment when the chance to destroy the enemy's fleet would have driven from me all thought of the conjoint strategy of the campaign as a whole.

I could not have helped it.

And I have never ceased to mourn the failure of the Count de Grasse to be as imprudent as I could not have helped being on that grandest of all occasions.

You will by no means infer from these cursory observations that I fail to appreciate, within my limited capacity, the grandeur of the tactical combinations, the skill of the intricate manoeuvres, and the far-sighted, long-thought out demonstrations by which the Count de Toulouse drove Rooke out of the Mediterranean in August 1704 with no more ado than the comparatively bloodless battle off Malaga; or the address with which La Galissonière repulsed Byng from Minorca in 1756 by a long-range battle of which the only notable casualty was the subsequent execution of Byng by his own government for the alleged crime of failing to destroy the fleet opposed to him! . . .

And yet, my dear Kersaint, one reflection persecutes me, to mar all my memories and baffle my admiration. This is the undeniable fact . . . that the ships and seamen of Graves, whom de Grasse permitted to escape from his clutches off the Capes of Chesapeake in October 1781, were left intact to discomfit de Grasse himself off Santa Lucia and Dominica in April 1782, under Rodney.

You know, of course, my dear Kersaint, that my own opportunities in naval warfare have been but few and feeble in comparison with such as I have mentioned. But I do not doubt your ready agreement with me if I say that the hostile ships and commanders that I have thus far enjoyed the opportunity of meeting did not give anyone much trouble thereafter. True, this has been on a small scale; but that was no fault of mine. I did my best with the weapons given to me.

The rules of conduct, the maxims of action, and the tactical instincts that serve to gain small victories may always be expanded into the winning of great ones with suitable opportunity; because in human affairs the sources of success are ever to be found in the fountains of quick resolve and swift stroke; and it seems to be a law inflexible and inexorable that he who will not risk cannot win.

ROBERT JACKSON:

FROM

A View of the Formation, Discipline and Economy of Armies

Robert Jackson was born in Scotland in 1750 and trained in medicine at Edinburgh—with vacations as a medical aide on a Greenland whaler. He then practiced medicine in the West Indies, joined the British army in New York after the outbreak of the Revolution as an assistant surgeon, and was captured. He returned to England on parole in 1782 and then wandered around Europe before rejoining the army in 1794. He served in Europe and the West Indies, revisited the United States, and was appointed inspector-general of hospitals at the beginning of the Peninsular War.

He published a report on maladministration in the army medical services, his own medical qualifications were called into question, and he assaulted one of his opponents —for which he was jailed for six months. After release he again took up a medical appointment in the West Indies and returned to England in 1815. Thereafter little is known about his career but he spent some time traveling in Europe, studying local conditions. He died in Scotland in 1827.

Jackson's view of war, as an army doctor, spans the American War of Independence and the Napoleonic Wars, in which he served.

Though not daring in close combat, they were not without courage. It was a courage of circumstance, the direct combat: front to front, was supported with resolution, the retrograde was precipitate when the flanks were turned, when the design of turning them was discovered, or when a front attack was threatened by the bayonet. This seemed to the writer to be the leading feature of the American military character during the revolutionary war; and, as it is in some measure a feature of circumstance, it is reasonable to believe that it resulted from habits engendered by mode of life.

The value of the American people as soldiers consists in skill in the use of fire-arms. That skill, it is presumed, arises from the practice of firing at birds and wild beasts in the rivers, ponds, and woods, of an extensive continent. Accustomed to circumvent, and to shoot from behind cover, the Americans were themselves afraid of being circumvented; and, impressed perhaps with the idea of circumvention, they moved off precipitately at the appearance of suspicious manoeuvres being practised against them: they had not, as a soldier ought to have, a face for flank and rear. The prey which the Americans were accustomed to pursue being a timid prey—to be entrapped rather than combatted by force, courage to face the enemy boldly was not acquired by the exercise of hunting: it was rather perhaps diminished by the habit of caution engendered by the practice of circumvention.

If the military merit of the American people, as it appeared during the revolutionary war, be estimated fairly, it does not stand high even in partizan war. The Americans were soldiers from necessity—not from genius or inclination. They did not proceed to the combat with a mind inflamed with ideas of national glory. They had little of military enterprize in the constitution originally; and they made little scientific progress in the military art during the continuance of the contest. They advanced boldly to action in several instances; they maintained no combat obstinately. The cover of a bank, a tree, or a fence, was necessary to give them confidence to look at their antagonist. They exercised the firelock with effect while they were under cover; they retired when the enemy approached near, that is, they split and squandered, according to the cant phrase, to rally at an assigned point in the rear.

If the attainment of superiority in the actual conflict of battle be the object of military training, the temper and energy of individuals ought, in the just reason of things, to be estimated so as to be known correctly to the full extent of their value. The exact order of external uniformity, according to which separate parts are arranged in the military fabric in the present time, is only a secondary object in the true meaning of things. Correspondence in power, not uniformity in the *coup d'oeil,* is the base of true military organization.

As it is in the temper of the parts, not in the uniformity of the *coup d'oeil* that the value of the military instrument consists, it is, or ought to be, the main object of the tactician, as frequently said, to arrange the parts in the ranks according to power and temper, rather than according to size and external resemblance. But it happens here, as it happens in many other things, that the ingenuity, or rather the presumption, of man counteracts his own design. Ignorant, or regardless of internal relations, he acts on the information of the eye, and thus gives a garb of order and dressing to the materials of the fabric which, as not resting on the true base, detracts from union, vigour, and consistency in the execution of function.

Hence it is that military education becomes vain, the effect comparatively void, or the reverse of good. Unless order be engrafted on the properties of the material with such care and discernment, that no part of the constitutional power and native spirit be marred or shackled by the artificial arrangement, the instinctive sagacity of the barbarian prevails over the science of the refined tactician. The fact is illustrated by the military history of semi-barbarous nations; who, though inferior in military arrangement, in the exterior forms of discipline, and greatly inferior in arms and military apparatus, not unfrequently defeat the armies of scientific, polished, and refined masters in the art of war.

The examples are numerous in the history of mankind; and even in recent times, the untaught peasantry of the poorer cantons of Switzerland, and of some part of Tyrol, gave more trouble to the troops of France than the regular armies of the great monarchs, which were exact in their movements as a machine of mechanical construction. Great Britain herself can speak to the fact. She sustained greater injury to her military reputation by the people of the town and district of Buenos Ayres and New Orleans, than from all the regular armies she encountered in the field during the late war.

The energy of spirit which leads to military enterprize is a quality of the early stage of society. It vanishes from nations in proportion as they become polished and refined; at least, it is not supported in a progressive course, unless by scientific study and a judicious application of such causes as, acting on human organism, maintain the machine in a state of activity to a forward point prominent in the view of all.

The exercises with the firelock, or common drillings of the European infantry, are not of a nature to interest the simple soldier. The purpose of them, as connected with utility, is not fully comprehended by him. He goes to the field as an automaton, to act and to be acted upon by mechanical powers, ignorant of the principle on which he acts, and the purpose for which he is constrained to act.

The mind is not interested by routine forms of duty; and, as it is important to success that the mind should be interested, it is useful, or may be supposed to be useful, to endeavour to give a new cast, consequently a new force of impression, to military exercises and military forms of evolution, without changing the principles of such practices as are laid on a basis of truth. New modes of military exercise interest the individual by their novelty; they even not infrequently communicate an animating energy to the arm of the actor, which goes beyond the limits of ordinary calculation: they seldom fail to intimidate the enemy as striking him by surprise. If this be so, it belongs to military genius to change the appearances of things, with a view to animate one part and to intimidate another. But, while this is done, especial care is to be taken that the fundamental principles of military tactic be not rashly violated.

The *Shrapnel* shell, as a means of extending the range of missile force, is an invention of science; and it may be considered as an important one in modifying the character of a military action. The *Congreve* rocket may surprise the inexperienced: it is a child's plaything in the field, rather than an instrument in war: it may be employed with advantage in sieges. The Polish lance, with which hussars have lately been armed, has had advantage on some occasions as an arm of offence; but it is chiefly to novelty that the unexpected effect is to be ascribed.

The broad-sword and target of the Scotch Highlander is perhaps inferior, in a correct estimate of the power of weapons, to the firelock and bayonet; it was notwithstanding formidable, and made a striking impression on British troops in the year 1745. The British soldier was armed in the year 1745 with the firelock and bayonet. He was a trained soldier, and moreover a soldier not unacquainted with the practice of war. The Highlander was rude, and unskilled in military tactic. If he carried a carbine into the field, he did not much rely on it. His chief trust was in the broad-sword. It was his national arm, and it was to him a talisman which gave confidence, even an idea of invincibility. With this arm and armour he discomfited the experienced troops of Great Britain, presumptively through surprise at the unknown mode of attack.

The Highlanders who fought on the continent of Europe, and in America in the war 1756, seemed to have acted on the French by a similar form of impression, as they had acted on the British at Prestonpans and Falkirk. Even so late as the American revolutionary war, the Highlanders, probably from the impression which the peculiarity of dress &c., made upon the peasant militia, were more dreaded than other British soldiers.

It is sufficiently proved in history that rude and semi-barbarous nations, ill armed and with little of what is called discipline, often discomfit the systematic armies of scientific tacticians and accomplished generals.

IV.
THE
REVOLUTION

What a change from 1785 to 1824! In two
thousand years of recorded history, so
sharp a revolution in the customs, ideas
and beliefs has never occurred before.

—STENDHAL

HORATIO NELSON:

FROM

The Trafalgar Memorandum;

FROM

The Diary

Horatio Nelson was born in Norfolk in 1758, the son of a parson. He joined the navy as a frail twelve-year-old midshipman and from then on his career became both heroic and, despite some setbacks and disappointments, triumphant. Dying in the hour of his last and greatest victory, at Trafalgar in 1805, he left England with over a century of undisputed maritime supremacy—and with his own impressively commemorated legend.

FROM *The Trafalgar Memorandum*

Victory off Cadiz,
9th October 1805

Thinking it almost impossible to bring a Fleet of forty Sail of the Line into a Line of Battle in variable winds, thick weather and other circumstances which must occur, without such a loss of time that the opportunity would probably be lost of bringing the Enemy to Battle in such a manner as to make the business decisive.

I have therefore made up my mind to keep the fleet in that position of sailing with the exception of the First and Second in Command that the order of sailing is to be the Order of Battle, placing the fleet in two Lines of Sixteen ships each with an advanced Squadron of Eight of the fastest sailing Two decked ships which will always make if wanted a Line of Twenty Four Sail, on which ever Line the Commander in Chief may direct.

The Second in Command will after my intentions are made known to him have the entire direction of His Line to make the attack upon the Enemy and to follow up the Blow until they are captured or destroyed.

The whole impression of the British must be, to overpower from two or three Ships ahead of their Commander in Chief, supposed to be in the Centre, to the Rear of their fleet. I will suppose twenty Sail of the Enemy's Line to be untouched. It must be some time before they could perform a Manoeuvre to bring their force compact to attack any part of the British fleet engaged, or to succour their own ships which indeed would be impossible, without mixing with the ships engaged. Something must be left to chance, nothing is sure in a sea fight beyond all others, shot will carry away the mast and yards of friends as well as foes, but I look with confidence to a victory before the van of the Enemy could succour their Rear and then that the British Fleet would most of them be ready to receive their Twenty Sail of the Line or to pursue them should they endeavour to make off.

If the Van of the Enemy tacks, the captured Ships must run to Leeward of the British Fleet, if the Enemy wears, the British must place themselves between the Enemy and the captured and disabled British Ships and should the enemy close I have no fear as to the result.

The Second in Command will in all possible things direct the Movements of his Line by keeping them as compact as the nature of the circumstances will admit. Captains are to look to their particular Line as their rallying point. But in case signals can neither be seen or perfectly understood no Captain can do very wrong if he places his Ship alongside that of an Enemy. . . .

. . . Should the Enemy wear together or bear up and sail Large still the Twelve Ships composing in the first position the Enemy's Rear are to be the Object of Attack of the Lee Line unless otherwise directed from the Commander in Chief which is scarcely to be expected as the entire management of the Lee Line after the intentions of the Commander in Chief are signified is intended to be left to the Judgement of the Admiral Commanding that Line.

The remainder of the Enemy's fleet 35 Sail are to be left to the

Management of the Commander in Chief who will endeavour to take care that the Movements of the Second in Command are as little interrupted as is possible.

FROM *The Diary, 21st October, 1805*

At daylight saw the Enemy's Combined Fleet from East to E.S.E.; bore away; made the signal for Order of Sailing, and to Prepare for Battle; the Enemy with their heads to the Southward; and at Seven the Enemy wearing in succession.

May the Great God whom I worship grant to my Country, and for the benefit of Europe in general, a great and glorious Victory; and may no misconduct in any one tarnish it; and may humanity after Victory be the predominant feature in the British Fleet. For myself individually, I commit my life to Him who made me, and may his blessing light upon my endeavours for serving my Country faithfully. To Him I resign myself and the just cause which is entrusted to me to defend. Amen. Amen. Amen.

NAPOLEON BONAPARTE:

FROM

Maxims;

FROM

Memoirs

"A man will arise, perhaps one who was hitherto lost in the obscurity of the crowd. . . . That man will seize hold of opinions, of opportunity, of fortune, and will say to the great man of theories what the practical architect, who addressed the Athenians, said to the oratorical architect: "All that my rivals tell you, I will carry out!"

—GUIBERT

Napoleon Bonaparte was born in Corsica in 1769 and joined the French army in 1785. With the revolutionaries he captured Toulon, suppressed a Royalist uprising in Paris, and was appointed commander of the Army of the Interior by the Directory. After a successful campaign in Italy, culminating in his defeat of the Austrians at Rivoli, he seized power in 1799. There followed a period of lasting reforms at home. The French frontiers were extended to the Rhine and the Alps with the victory of Marengo in 1800, before the Peace of Amiens. The resumption of hostilities was followed by the frustration of a projected invasion of Britain, but also by triumphs on the Continent, and the victories of Ulm, Austerlitz, Jena, Friedland, and Wagram. Napoleon made himself emperor in 1804 and master of much of Europe through his dependents. Then the disas-

trous Russian campaign of 1812, the British successes in the Spanish Peninsular War, and the Allied victory at Leipzig in 1813 brought about Napoleon's downfall in 1814. Returning from Elba the following year he was finally defeated at Waterloo and exiled to St. Helena where he died in 1821.

FROM *Maxims*

A general-in-chief cannot exonerate himself from responsibility for his faults by pleading an order of his sovereign or the minister, when the individual from whom it proceeds is at a distance from the field of operations, and but partially, or not at all, acquainted with the actual condition of things. Hence it follows that every general-in-chief who undertakes to execute a plan which he knows is bad, is culpable. He should communicate his reasons, insist on a change of plan, and finally resign his commission rather than become the instrument of his army's ruin.

Every general-in-chief who, in consequence of orders from his superiors, gives battle with the certainty of defeat is equally culpable.

In this latter case, he should refuse to obey; for an order requires passive obedience only when it is issued by a superior who is present at the seat of war. As the superior is then familiar with the state of affairs, he can listen to the objections and make the necessary explanations to the officer who is to execute the command.

But suppose a general-in-chief were to receive from his sovereign an order to give battle with the injunction to yield the victory to his adversary and permit himself to be beaten. Would he be bound to obey? No! If the general comprehended the futility of so strange an order, he ought to execute it; but, if not; he should refuse to obey.

Commanders-in-chief are to be guided by their own experience or genius. Tactics, evolution and the science of the engineer and the artillery officer may be learned from treatises, but generalship is acquired only by experience and the study of the campaigns of all great captains. Gustavus Adolphus, Turenne and Frederick, as also Alexander, Hannibal and Caesar have all acted on the same principles. To keep your forces united, to be vulnerable at no point, to bear down with rapidity upon important points—these are the principles which insure victory.

FROM *Memoirs*

War by land generally destroys more men than maritime war, being more perilous. The sailor, in a squadron, fights only once in a campaign; the soldier fights daily. The sailor, whatever may be the fatigues and dangers attached to his element, suffers much less than the soldier; he never endures hunger or thirst, he has always with him his lodging, his kitchen, his hospital, and medical stores. The naval armies, in the service of France and England, where cleanliness is preserved by discipline, and experience has taught all the measures proper to be adopted for the preservation of health, are less subject to sickness than land armies. Besides the dangers of battle, the sailor has to encounter those of storms; but art has so materially diminished the latter, that they cannot be compared to those which occur by land, such as popular insurrections, assassinations, and surprises by the enemy's light troops.

A general who is commander-in-chief of a naval army, and a general who is commander-in-chief of a land army, are men who stand in need of different qualities. The qualities adapted to the command of a land army are born with us, whilst those which are necessary for commanding a naval army can only be acquired by experience.

Alexander and Condé were able to command at a very early age; the art of war by land is an art of genius and inspiration; but neither Alexander nor Condé, at the age of twenty-two years, could have commanded a naval army. In the latter, nothing is genius or inspiration, but all is positive and matter of experience. The marine general needs but one science, that of navigation. The commander by land requires many, or a talent equivalent to all, that of profiting by experience and knowledge of every kind.

A marine general has nothing to guess; he knows where his enemy is, and knows his strength. A land general never knows anything with certainty, never sees his enemy plainly, nor knows positively where he is. When the armies are facing each other, the slightest accident of the ground, the least wood, may hide a party of the hostile army. The most experienced eye cannot be certain whether it sees the whole of the enemy's army, or only three-fourths of it. It is by the eyes of the mind, by the combination of all reasoning, by a sort of inspiration, that the land general sees, commands, and judges. The marine general requires nothing but an experienced eye; nothing relating to the enemy's strength is concealed from him.

What creates great difficulty in the profession of the land commander, is the necessity of feeding so many men and animals; if he allows himself to be guided by the commissaries, he will never stir, and his expeditions will fail. The naval commander is never confined; he carries everything with him. A naval commander has no reconnoitring to perform, no ground to examine, no field of battle to study; Indian Ocean, Atlantic, or Channel, still it is a liquid plain. The most skilful can have no other advantage over the least experienced, than what arises from his knowledge of the winds which prevail in particular seas, from his foresight of those which will prevail there, or from his acquaintance with the signs of the atmosphere; qualities which are acquired by experience, and experience only.

The general commanding by land never knows the field of battle on which he is to operate. His *coup d'oeil* is one of inspiration, he has no positive data. The data from which a knowledge of the localities must be gained, are so contingent, that scarcely anything can be learnt from experience. It is a facility of instantly seizing all the relations of different grounds, according to the nature of the country; in short, it is a gift called *coup d'oeil militaire,* which great generals have received from nature. Nevertheless, the observations which may be made on topographical maps, and the facilities arising from education and the habit of reading such maps, may afford some assistance.

A naval commander-in-chief depends more on the captains of his ships, than a military commander-in-chief on his generals. The latter has the power of taking on himself the direct command of the troops, of moving to every point, and of remedying the false movements by others. The personal influence of the naval commander is confined to the men on board his own ship; the smoke prevents the signals from being seen. The winds change, or may not be the same throughout the space occupied by his line. Of all arts, then, this is the one in which the subalterns have the most to take upon themselves.

Our naval defeats are to be attributed to three causes: first, to irresolution and want of energy in the commanders-in-chief; secondly, to errors in tactics; thirdly, to want of experience and nautical knowledge in the captains of ships, and to the opinion these officers maintain that they ought only to act according to signals. The action off Ushant, those during the Revolution in the Ocean, and those in the Mediterranean in 1793 and 1794, were all lost through these different causes. Admiral Villaret, though personally brave, was wanting in strength of mind, and was not even attached to the cause for which he fought. Martin was a good seaman, but a man of little resolution. They were, moreover, both influenced by the Representatives of the People, who, possessing no experience, sanctioned erroneous operations.

The principle of making no movement, except according to signal from the admiral, is the more erroneous, because it is always in the power of the captain of a ship to find reasons in justification of his failure to execute the signals made to him. In all the sciences necessary to war, theory is useful for giving general ideas which form the mind; but their strict execution is always dangerous; they are only axes by which curves are to be traced. Besides rules themselves compel one to reason in order to discover whether they ought to be departed from.

Although often superior in force to the English, we never knew how to attack them, and we allowed their squadrons to escape whilst we were wasting time in useless manoeuvres. The first law of maritime tactics ought to be, that as soon as the admiral has made the signal that he means to attack, every captain should make the necessary movements for attacking one of the enemy's ships taking part in the action, and supporting his neighbours.

This was latterly the principle of English tactics. Had it been adopted in France, Admiral Villeneuve would not have thought himself blameless at Aboukir, for remaining inactive with five or six ships, that is to say, with half the squadron, for twenty-four hours, whilst the enemy was overpowering the other wing.

The French navy is called on to acquire a superiority over the English. The French understand building better than their rivals, and French ships, the English themselves admit, are better than theirs. The guns are superior in calibre to those of the English by one-fourth. These are two great advantages.

The English are superior in discipline. The Toulon and Scheldt squadrons had adopted the same practice and customs as the English, and were attempting as severe a discipline, with the difference belonging to the character of the two nations. The English discipline is perfectly slavish; it is patron and serf. It is only kept up by the influence of the most dreadful terror. Such a state of things would degrade and debase the French character, which requires a paternal kind of discipline, more founded on honour and sentiment.

—SOMERSET DE CHAIR
(translator)

DUKE OF WELLINGTON:

FROM

Despatches

"It makes me *burn* to have been a soldier."
—CARDINAL NEWMAN, *on reading the* Despatches

Arthur Wellesley was born in Dublin in 1769 and attended a French military academy after schooling at Eton. He joined a Highland regiment, gained rapid promotion through family money and meanwhile sat in the Irish Parliament. As a commander in India he won victories in the Maratha War and, on his return, took the political post of Irish Secretary. He defeated a Danish force in an expedition against Copenhagen and was sent to Portugal on the outbreak of the Peninsular War. His defensive strategy against much superior forces was followed by a series of victories at Cuidad Rodruigo, Badajoz, Salamanca, and Vittoria. Finally entering France, he defeated the French at Toulouse before participating in the peace settlement. On Napoleon's return, he took command and defeated the French at Waterloo in 1815. As Duke of Wellington he returned to a political career and became prime minister in 1828, passing the Catholic Emancipation bill. He resigned, amid riots, in 1830 when the Great Reform bill was passed by the new government with his belated support. For many years he remained in public life and was an inactive commander in chief until his death in 1852.

"Really when I reflect upon the character and the attainment of some of the General officers of this army, and consider that these are the persons on whom I am to rely to lead columns against the French generals, and who are to carry my instructions into execution, I tremble; and, as Lord Chesterfield said of the Generals of his day, 'I only hope that when the enemy reads the list of their names he trembles as I do.'"

—WELLINGTON, 1810

Letter to Sydenham from Portugal, 1811

Freneda, 7th Dec. 1811

I am writing without much detailed knowledge upon this subject; at the same time, I write in hopes that my reflections and suggestions may be of use.

In regard to other points, I hope that the blow will not be struck too soon; and that the Sovereigns of Europe, and all who are determined to resist Buonaparte, will enter into the plan with a determination to persevere until they put an end to the system of making war as a financial resource; that they won't proceed according to the old plan of sacrificing a part to save the remainder, but will one and all persevere to the last, and either save all or lose all; with an entire conviction that the *remainder,* as it is called, will, in the course of events, be taken from them, if they should cease to resist. I have given the Duke of Brunswick a lesson on this subject, which will not be useless if he has any communication with the Continent.

But the principal point on which I wished to write to you is the disposal of this army, supposing that there should be a general breeze in Europe. I think that you have miscalculated the means and resources of France in men, and mistaken the objects of the French government in imagining that, under those circumstances, Buonaparte will be obliged or inclined to withdraw his army from Spain. He will not even reduce it considerably, but he will only not reinforce it. If I am right, the British army cannot be so advantageously employed as in the Peninsula. Of that, I trust, there is no doubt. If the British army is not employed in the Peninsula, that part of the world would soon be conquered; and the army which would have achieved its conquest, reinforced by the levies in the Peninsula, would reduce to subjugation the rest of the world.

But that is not exactly the view which you have taken of the subject. You appear to think it probable that Buonaparte would be inclined or obliged to withdraw from the Peninsula; and you ask, what would I do in that case? I answer, attack the most vulnerable frontier of France, that of the Pyrenees. Oblige the French to maintain in that quarter 200,000 men for their defence; touch them vitally there, when it will certainly be impossible to touch them elsewhere, and form the nations of the Peninsula into soldiers, who would be allies of Great Britain for centuries.

I acknowledge that this task would not be an easy one. But the difficulties are not insurmountable; and I think it is possible, with our maritime resources, to form and maintain an army in the Pyrenees.

But there is another view of this question; and that is, what shall we do with this army if we don't employ it in the Peninsula? Government must first begin by understanding clearly that they cannot have this army in any part of the north of Europe, or in the Adriatic, in a state fit for service in less than 6 months after the resolution to alter its destination shall be passed in the Cabinet. Who can foresee the events of 6 months in these days? Who can depend upon his own foresight?

But there is another consideration to be adverted to. I have a gross 50,000 men, but I have never had 35,000 fit for duty in the field. But I will suppose that government could reckon upon 40,000 British troops. These could not act separately in Germany, or in the north of Europe: they must be attached to some other army, of which no power in Europe would give the British government the command. Could the British government leave its army at the disposal of Prussia, Russia, or Austria, or of the insurgents in the north of Europe? Then comes the question of supplies. The British army must be well supplied, and might and would be well supplied in the north of Europe. But it would soon be found that the foreign army to which it should be attached must be supplied likewise. Would the British government be prepared to defray the expense of both? If they could not, the British army would starve equally with the foreign army; and what would the British public say to this?

I believe that the Court and head quarters of the army are very desirous of changing the seat of our operations, for reasons into which it is not now necessary to enter. Not so the public. But our ministers may depend upon it that they cannot establish any where such a system as they have here; that they cannot any where keep in check so large a proportion of Buonaparte's army, with such small comparative British means; that they cannot any where be principals, and carry on the war upon their own responsibility, at so cheap a rate of men and means as they can here; that no seat of operations holds out such prospects of success, whatever may happen elsewhere, even for the attainment of those objects which would be in view in transferring the seat of the war to the north of Europe.

Letter to Earl Bathurst from France, 1813

St. Jean de Luz, 21st Nov., 1813

I enclose you an original address which has been presented to me by the constituted authorities and notables here (which I hope your Lordship will do me the favor not to make public), which will show the strong sentiment here respecting the war; the same prevailed at St. Pé, and I hear of the same opinions in all parts of the country.

I have not myself heard any opinion in favor of the House of Bourbon. The opinion stated to me upon that point is, that 20 years have elapsed since the Princes of that House have quitted France; that they are equally, if not more, unknown to France than the Princes of any other Royal House in Europe; but that the allies ought to agree to propose a Sovereign to France instead of Napoleon, who must be got rid of, if it is hoped or intended that Europe should ever enjoy peace; and that it was not material whether it was of the House of Bourbon or of any other Royal Family.

I have taken measures to open correspondence with the interior, by which I hope to know what passes, and the sentiments of the people, and I will take care to keep your Lordship acquainted with all that I may learn. In the mean time, I am convinced more than ever that Napoleon's power stands upon corruption, that he has no adherents in France but the principal officers of his army, and the *employés civils* of the government, and possibly some of the new proprietors; but even these last I consider doubtful.

Notwithstanding this state of things, I recommend to your Lordship to make peace with him if you can acquire all the objects which you have a right to expect. All the powers of Europe require peace possibly more than France, and it would not do to found a new system of war upon the speculations of any individual on what he sees and learns in one corner of France. If Buonaparte becomes moderate, he is probably as good a Sovereign as we can desire in France; if he does not, we shall have another war in a few years; but if my speculations are well founded, we shall have all France against him; time will have been given for the supposed disaffection to his government to produce its effect; his diminished resources will have decreased his means of corruption, and it may be hoped that he will be engaged singlehanded against insurgent France and all Europe.

There is another view of this subject, however, and that is, the continuance of the existing war, and the line to be adopted in that case. At the present moment it is quite impossible for me to move at all: although the army was never in such health, heart, and condition as at present, and it is probably the most complete machine for its numbers now existing in Europe, the rain has so completely destroyed the roads that I cannot move; and, at all events, it is desirable, before I go farther forward, that I should know what the allies propose to do in the winter, which I conclude I shall learn from your Lordship as soon as the King's government shall be made acquainted with their intentions by the King's diplomatic servants abroad. As I shall move forward, whether in the winter or the spring, I can acquire and ascertain more fully the sentiments of the people, and the government can either empower me to decide to raise the Bourbon standard, or can decide the question hereafter themselves, after they shall have all the information before them which I can send them of the sentiments and wishes of the people.

I can only tell you that, if I were a Prince of the House of Bourbon, nothing should prevent me from now coming forward, not in a good house in London, but in the field in France; and if Great Britain would stand by him, I am certain he would succeed. This success would be much more certain in a month or more hence, when Napoleon commences to carry into execution the oppressive measures which he must adopt in order to try to retrieve his fortunes.

I must tell your Lordship, however, that our success, and every thing, depends upon our moderation and justice, and upon the good conduct and discipline of our troops. Hitherto these have behaved well, and there appears a new spirit among the officers, which I hope will continue, to keep the troops in order. But I despair of the Spaniards. They are in so miserable a state, that it is really hardly fair to expect that they will refrain from plundering a beautiful country, into which they enter as conquerors; particularly, adverting to the miseries which their own country has suffered from its invaders. I cannot, therefore, venture to bring them back into France, unless I can feed and pay them; and the official letter which will go to your Lordship by this post will show you the state of our finances, and our prospects. If I could now bring forward 20,000 good Spaniards, paid and fed, I should have Bayonne. If I could bring forward 40,000, I don't know where I should stop. Now I have both the 20,000 and the 40,000 at my command, upon this frontier, but I cannot venture to bring forward any for want of means of paying and supporting them. Without pay and food, they must plunder; and if they plunder, they will ruin us all.

I think I can make an arrangement of the subsidy to cover the expense of 20,000 Spaniards; but all these arrangements are easily settled, if we could get the money. Where we are to get the money, excepting from England, it is impossible for me to devise; as the patriotic gentlemen at Lisbon, now that they can buy no commissariat debts, will give us no money, or very little, for the drafts on the Treasury, and the yellow fever has put a stop to the communication with Cadiz and Gibraltar; and if we had millions at all three, we could not get a shilling for want of ships to bring it.

PRIVATE WHEELER:

FROM

Letter after Waterloo

Little is known about Private Wheeler apart from what he tells of his service in the King's Own Yorkshire Light Infantry during the Napoleonic Wars. On his retirement from the army with a small pension, he lived quietly for a number of years in the north of England. Wheeler was an admirer of both Wellington and Napoleon.

Cambrai 25th June 1815

On the morning of the 24th inst we marched on Cambray, about a league from the town we fell in with some cavalry picquets. After passing them we soon came in sight of the town, saw the tricolour flag flying on the citadel. . . .

We now pushed on to the works, near the gate, got into the trenches, fixed our ladders, and was soon in possession of the top of the wall. The opposition was trifling, the regular soldiers fled to the citadel, and the shopkeepers to their shops. We soon got possession of the gate and let in the remainder of the brigade, formed and advanced to the great square. We were, as was usual, received by the people with vivas, many of whom had forgot to wash the powder off their lips caused by biting off the cartridges when they were firing on us from the wall. The remainder of the division entered the town at the same time on the opposite side.

Picquets were established at the citadel, and about dusk the remainder of the division were marched out of the town and encamped. We had picked up some money in the town, or more properly speaking we had made the people hand it over to us to save the trouble of taking it from them, so we were enabled to provide ourselves with what made us comfortable. . . .

On the 25th we halted and His pottle belly Majesty, Louis 18th, marched into the loyal town of Cambray. . . . No doubt the papers will inform you how Louis 18th entered the loyal city of Cambray, how his loyal subjects welcomed their beloved king, how the best of monarchs wept over the sufferings of his beloved people, how the Citadel surrendered with acclamations of joy to the best of kings, and how his most Christian Majesty effected all this without being accompanied by a single soldier. But the papers will not inform you that the 4th Division and a brigade of Hanovarian Huzzars (red) were in readiness within half a mile of this faithful city, and if the loyal citizens had insulted their king, how it was very probable we should have bayoneted every Frenchman in the place. The people well knew this, and this will account for the sudden change in their loyalty or allegiance from their Idol Napoleon (properly named the Great) to an old bloated poltroon, the Sir John Falstaff of France.

ARMAND DE CAULAINCOURT:

FROM

Memoirs

Armand de Caulaincourt, Duc de Vicence, was born in 1773 and joined the French army in 1795. He was colonel of a cavalry regiment at the battle of Hohenlinden and was sent on a diplomatic mission to Russia in 1802. He was involved in countersubversive activities on his return and then returned to Russia as ambassador from 1807 to 1811. As Grand Equerry he accompanied Napoleon on his retreat from Moscow, as he had in earlier battles, and after the battle of Leipzig he was appointed foreign minister. He negotiated the peace, which sent Napoleon to Elba, but rejoined the emperor as foreign minister during the Hundred Days. Thereafter he lived in retirement until his death in 1827.

The Memoirs, *based on the author's close political and military association with Napoleon, including the long ride back together from Moscow when they discussed war, were not published until the 1930s.*

According to the Emperor the presence of the English Army was the greatest obstacle to the pacification of Spain, but he would rather see it in that country than be threatened with it at any moment —in Brittany or Italy, or anywhere, in fact, where the coast was accessible. As it was, he knew where to look for the English; while if they were not occupied there he would be forced to prepare for them, and hold himself ready for defence against them, at every point. And that would use up many more troops, give him much more anxiety, and possibly do him much more damage.

"If 30,000 English landed in Belgium," he said to me, "or in the Pas-de-Calais, and requisitioned supplies from three hundred villages —if they were to go and burn the chateau of Caulaincourt—they would do us much more harm than by forcing me to maintain an army in Spain. You would make a much worse outcry, my good Master of the Horse! You would complain much more loudly than you do when you say that I aim at universal monarchy! The English are playing into my hands. If the Ministry were in my pay they could not act in a way more favourable to me. You must take good care not to repeat the ideas I express to you; for if the idea entered their heads to make expeditions against my coasts, now at one point and now at another: to re-embark as soon as forces were collected to fight them, and go at once to threaten some other point—the situation would be unsupportable."

"As it is," he added, "the war in Spain costs me no more than any other war, or any other compulsory defence against the English. So long as peace is not made with that Power, there is not much difference in cost between the present state of affairs in Spain and an ordinary state of war with England. In view of the great length of Spain's coast-line, with the situation as it is at present we must limit ourselves to keeping the English under observation—unless, indeed, they should march into the interior and a highly favourable opportunity arise for giving battle; for if we forced them to re-embark at one point, since they would always be sure of finding auxiliaries, they would disembark again at another.

The Marshals and Generals who have been left to look after themselves in Spain might have done better, but they will not come to an agreement. There has never been any unity in their operations. They detest each other to such an extent that they would be in despair if one thought he had made a movement that might yield credit to another. Accordingly there is nothing to be done except hold the country and try to pacify it until I can myself put some vigour into the operations there. Soult has ability: but no one will take orders. Every General wants to be independent, so as to play the viceroy in his own province.

"In Wellington," he added, "my Generals have encountered an opponent superior to some of them. Moreover, they have made the mistakes of a schoolboy. Marmont shows a really high quality of judgment and logic in discussing war, but is not even moderately able in action. In fact, our momentary reverses in that war, which delight the city of London, have little effect on the general course of affairs—and cannot indeed have any real importance, as I can change the face of affairs when I please.

"Events at present," he said, "are giving Wellington a reputation; but in war men may lose in a day what they have spent years in building up. As to the outlet for English trade which the war has created in the Spanish colonies, I admit that is certainly unfortunate as within two years those outlets may counterbalance our prohibition of imports on the Continent."

The Emperor saw, in the separation of these colonies from their metropolis, an important point which would change the politics of the world, which would give new strength to America, and in less than ten years would threaten the power of the English—which would be a compensation. He did not question that Mexico, and all the major Spanish possessions overseas, would declare their independence and form one or two States under a form of government which would force them, in their own interests, to become auxiliaries of the United States.

"It marks a new era," he said. "It will lead to the independence of all other colonies."

The changes that would arise from this development he regarded as the most important of the century, since they would alter the balance of commercial interests and, in consequence, alter the policy of the different Governments.

"All the colonies," he said, "will imitate the United States. The colonials grow tired of obeying a Government which seems foreign to them because it subordinates them to its own local interests, interests which it cannot sacrifice to theirs. As soon as they feel strong enough to resist, the colonies want to shake off the yoke of those who created them. One's country is where one lives; a man does not take long to forget that he or his father was born under another sky. Ambition achieves what self-interest has begun. They want to have a standing of their own and then the yoke is soon thrown off."

I spoke to the Emperor of the moral effect which the resistance of the Spanish nation was having on people in general, suggesting to him that he was mistaken in attaching no importance to the example they were setting. I reminded him of the remark of the Tsar Alexander, which had struck me and which I had repeated to him on my return: "You have beaten the Spanish armies but you have not subdued the nation. The nation will raise other armies. The Spaniards, without any government, are setting a noble example to other nations. They are teaching the sovereigns what can be accomplished by perseverance in a just cause."

The Emperor treated as a joke what he called "the utterances of the prophet of the North."

Returning to affairs in Spain, the Emperor said:

"It is easy to pronounce judgment upon what is past: and easy to exalt as heroism what depends upon causes that are in truth hardly honourable. The heroism with which, in their hatred of France, they now credit the Spaniards arises simply out of the barbarous condition of that half-savage population and out of the superstitions to which the mistakes of our Generals have given new vigour. It is out of laziness, not out of heroism, that the Spanish peasants prefer the dangerous life of a smuggler or of a highwayman to the labours of cultivating the soil. The Spanish peasants have seized the opportunity of taking up this nomadic, smuggler's existence which is so suited to their taste and so much to the advantage of their poverty-stricken condition. There is nothing patriotic about that."

* * *

The Emperor occupied himself with the most minute details. He wanted everything to bear the imprint of his genius. He would send for me to receive his orders for headquarters, for the orderly officers, for his staff officers, for the letters, for the couriers, postal service, etc. The commanding officer of the Guard; the controller of the army commissariat; Larrey, the excellent surgeon-general, all were summoned at least once a day. Nothing escaped his solicitude. Indeed, his foresight might well be called by the name of solicitude, for no detail seemed too humble to receive his attention. Whatever might contribute to the success or well-being of his soldiers appeared to him worthy of daily care. Never can it be said of the Emperor that he was lulled into slumber by prosperity, for however great a victory he may have won, at the very moment that success was assured he occupied himself with as many precautions as he would have taken had it been a defeat.

Even when chasing the enemy helter-skelter before him, or in the heat of one of his greatest victories, no matter how weary the Emperor was he always had an eye for ground that could be held in the event of a reverse. In this respect he had an astonishing memory for localities. The topography of a country seemed to be modelled in relief in his head. Never did any man combine such a memory with a more creative genius. He seemed to extract men, horses and guns from the very bowels of the earth. The distinctive numbers of his regiments, his army service companies, his baggage battalions, were all classified in his brain most marvellously. His memory sufficed for everything. He knew where each one was, when it started, when it should arrive at its destination. His memory was more trustworthy than any staff musters and rolls, but this spirit of orderliness to the end that all should co-operate to achieve his purpose, that all should be created and organized with the final aim in view, did not go beyond that point.

All would have been well if the solution of the problems of the campaign could have been secured by gaining two or three battles: he was so completely master of his chessboard that he would certainly have won them. But his creative genius had no knowledge of conserving its forces. Always improvising, in a few days he would consume, exhaust and disorganize by the rapidity of his marches, the whole of what his genius had created. If a thirty-days' campaign did not produce the results of a year's fighting the greater part of his calculations were upset by the losses he suffered, for everything was done so rapidly and unexpectedly, the chiefs acting under him had so little experience, showed so little care and were, in addition so spoiled by former successes, that everything was disorganized, wasted and thrown away.

The Emperor's genius had proved itself in the achievement of such prodigious successes that to him was left the entire responsibility of winning a battle. It was sufficient to be on the spot in time for the action; after the victory had been won there was certain to be plenty of time to rest and reorganize, so no one cared very much what his losses had been or what he had had to abandon, for it was rare that the Emperor demanded an account. The prompt results of the Italian and Austrian campaigns and the resources those countries offered to the invader spoiled everyone, down to the less important commanders, for more rigorous warfare.

The habit of victory cost us dear when we got to Russia and even dearer when we were in retreat; the glorious habit of marching ever forward made us veritable schoolboys when it came to retreating. The Emperor was so used to having his troops at hand and was always so eager to take the offensive that the roads became hopelessly blocked and the columns inextricably confused. In this matter men and horses alike were reduced to a state of exhaustion.

Never was a retreat worse planned, or carried out with less discipline; never did convoys march so badly. Precautionary calculations and dispositions had no place in the arrangements that were made and it was to this lack of forethought that we owed a great part of our disaster. When it came to any retrograde movement the Emperor would take no decision until the very last moment, which was invariably too late. His reasoning powers were never able to gain the mastery over his repugnance to retreat, while his staff, who were far too much in the habit of not doing the slightest thing without the impulse from him who planned everything, took no steps whatever to organize affairs. Shaped and drilled into being no more than an obedient instrument, the staff could do nothing of itself for the general good.

The Emperor would not even agree to the most essential sacrifice to preserve what was undoubtedly indispensable. Throughout that long retreat from Russia he was as uncertain and as undecided on the last day as he had been on the first, although he was in no more doubt as to the imperative necessity of this retreat than anyone else. Constantly deluding himself with hopes of being able to call a halt and take up a position, he obstinately retained an immense amount of material that ultimately caused the loss of everything. He had a wholly incalculable antipathy for any thought or ideas about what he disliked. Fortune had so often smiled upon him that he could never bring himself to believe that she might prove fickle.

—George Libaire
(translator)

ANTOINE DE JOMINI:

FROM

Summary of the Art of War

Baron Antoine de Jomini was born in Switzerland in 1779 and, after working in a Paris bank, organized battalions in the Swiss army. His writings on tactics brought him fame and he was appointed aide-de-camp to Marshal Ney and a colonel by Napoleon. After accompanying Ney on the Jena and Eylau campaign and also to Spain, he resigned from the French service, but was recalled by Napoleon in 1810, at the age of twenty-eight, with the rank of general of brigade.

He became director of the historical section of the French general staff and then participated in the Russian campaign, acting as Ney's chief of staff. In 1813 he deserted, after largely winning the battle of Bautze but failing to receive promotion, and served as a lieutenant general in the Russian army for the rest of the war. He was present as aide-de-camp to the Emperor Alexander at the Battle of Leipzig.

He took part in the Congress of Vienna, became military tutor to the Russian Imperial family and accompanied his pupil, now the Emperor Nicholas I, on the Turkish campaign of 1828 with the rank of general in chief. He helped to found the Military Academy in Moscow in 1832. His most famous work, Précis de l'Art de la Guerre, *was prepared for the Tsarevich Alexander and published in 1837. He retired in 1848, though he returned to advise the emperor during the Crimean War in 1854, and died in Paris in 1869.*

One cannot deny to General Clausewitz great learning and a facile pen. But this pen, at times a little vagrant, is above all too pretentious for a didactic discussion, in which simplicity and clearness ought to come first. Besides that, the author shows himself by far too sceptical in point of military science; his first volume is but a declamation against all theory of war, whilst the two succeeding volumes, full of theoretic maxims, prove that the author believes in the efficacy of his own doctrines, if he does not believe in those of others. Of all theories on the art of war, the only reasonable one is that which, founded upon the study of military history, admits a certain number of regulating principles but leaves to natural genius the greatest part of the general conduct of a war without trammelling it with exclusive rules.

* * *

The love of conquest, however, was not the only motive with Napoleon: his personal position and his contest with England urged him to enterprises calculated to make him supreme. One might say his victories teach us what may be accomplished by activity, boldness and skill; his disasters, what might have been avoided by prudence.

* * *

As a soldier preferring loyal and chivalrous warfare to organized assassination if it be necessary to make a choice, I acknowledge that my prejudices are in favour of the good old times when the French and English Guards courteously invited each other to fire first—as at Fontenoy—preferring them to the frightful epoch when priests, women, and children through Spain plotted the murder of isolated soldiers.

* * *

The action of a cabinet in reference to the control of armies influences the boldness of their operations. A general whose genius and hands are tied by an Aulic council five hundred miles distant cannot be a match for one who has liberty of action, other things being equal.

If the skill of a general is one of the surest elements of victory, it will readily be seen that the judicious selection of generals is one of the most delicate points in the science of government and one of the most essential parts of the military policy of a state. Unfortunately, this choice is influenced by so many petty passions that chance, rank, age, favour, party spirit, or jealousy will have as much to do with it as the public interest and justice.

Superiority of armament may increase the chances of success in war. It does not of itself gain battles, but it is a great element of success.

The new inventions of the last twenty years seem to threaten a great revolution in army organization, armament and tactics. Strategy alone will remain unaltered, with its principles the same as under the Scipios and Caesars, Frederick and Napoleon, since they are independent of the nature of arms and the organization of the troops.

In times of peace the general staff should plan for all possible contingencies of war. Its archives should contain the historical details of the past, and all statistical, geographical, topographical and strategic treatises and papers for the present and future.

The financial condition of a nation is to be weighed among the chances of war. Still it would be dangerous to constantly attribute to this condition the importance attached to it by Frederick the Great in the history of his times. . . . If England has proved that money will procure soldiers and auxiliaries, France has shown that love of country and honour are equally productive and that, when necessary, war may be made to support war. . . . Still we must admit that a happy combination of wise military institutions, of patriotism, of well-regulated finances and of internal wealth and public credit imparts to a nation the greatest strength and makes it best capable of sustaining a long war.

* * *

One great principle underlies all the operations of war—a principle which must be followed in all good combinations. It is embraced in the following maxims:

1. To throw by strategic movements the mass of an army, successively, upon the decisive points of a theatre of war, and also upon the communications of the enemy as much as possible without compromising one's own.

2. To manoeuvre the engaged fractions of the hostile army with the bulk of one's forces.

3. On the battlefield, to throw the mass of the forces upon the decisive point, or upon that portion of the hostile line which it is of the first importance to overthrow.

4. To so arrange that these masses shall not only be thrown

upon the decisive point, but that they shall engage at the proper times and with ample energy.

Battles have been stated by some writers to be the chief and deciding features of war. This assertion is not strictly true, as armies have been destroyed by strategic operations without the occurrence of pitched battles, merely by a succession of inconsiderable affairs.

It is also true that a complete and decided victory may bring similar results even though there may have been no grand strategic combination. But it is the morale of the armies, as well as of nations, more than anything else, which makes victories and their results decisive. Clausewitz commits a grave error in asserting that a battle not characterised by a manoeuvre to turn the enemy cannot result in a complete victory.

At the battle of Zama, Hannibal in a few brief hours saw the fruits of twenty years of glory and success, vanish before his eyes, although Scipio never had a thought of turning his position. At Rivoli the turning-party was completely beaten. Nor was the manoeuvre more successful at Stochach in 1799 or at Austerlitz in 1805. I by no means intend to discourage the use of that manoeuvre, being on the contrary a constant advocate of it—but it is very important to know how to use it skillfully and opportunely. Moreover I am of the opinion that if it be a general's design to make himself master of his enemy's communications while at the same time holding his own, he should employ strategic rather than tactical combinations to accomplish it.

* * *

Posterity will regret, as the loss of an example to all future generations, that this immense undertaking (Napoleon's invasion of Britain) was not carried through, or at least attempted. Doubtless many brave men would have met their deaths, but were not those men mowed down more uselessly on the plains of Swabia, of Moravia, and of Castile, in the mountains of Portugal and the forests of Lithuania? What man would not glory in taking part in the greatest trial of skill and strength ever seen between two great nations?

At any rate posterity will find in the preparation made for this descent one of the most valuable lessons the present century has furnished for the study of soldiers and of statesmen. The labours of every kind performed on the coasts of France from 1803 to 1805 will be among the most remarkable monuments of the activity, foresight, and skill of Napoleon. It is recommended to the careful attention of young officers.

* * *

One of the surest ways of forming good combinations in war would be to order movements only after obtaining perfect information of the enemy's proceedings. In fact, how can any man say what he should do himself, if he is ignorant what his adversary is about? Even as it is unquestionably of the highest importance to gain this information, so it is a thing of the utmost difficulty, not to say impossibility. This is one of the chief causes of the great difference between the theory and the practice of war.

An attempt of another kind was made in 1794, at the battle of Fleurus, where General Jourdan made use of the services of a balloonist to observe and give notice of the movements of the Austrians. I am not aware that he found the method very useful, as it was not again used but it was claimed at the time that it assisted in gaining him the victory. Of this, however, I have great doubts.

It is probable that the difficulty of having a balloonist in readiness to make an ascension at the proper moment and of making careful observations upon what is going on below while floating at the mercy of the winds above, has led to the abandonment of this method of gaining information. By giving the balloon no great elevation, sending up with it an officer capable of forming correct opinions as to the enemy's movements, and perfecting a system of signals to be used in connection with the balloon, considerable advantages might be expected from its use. Sometimes the smoke of the battle and the difficulty of distinguishing the columns, that look like lilliputians, so as to know to which party they belong, will make the reports of the balloonists very unreliable. For example, a balloonist would have been greatly embarrassed in deciding, at the battle of Waterloo, whether it was Grouchy or Blucher who was seen coming up by the Saint Lambert road—but this uncertainty need not exist where the armies are not so much mixed.

I had ocular proof of the advantage to be derived from such observations when I was stationed in the spire of Gautsch, at the battle of Leipzig; and Prince Schwartzenberg's aide-de-camp, whom I had conducted to the same point, could not deny that it was at my solicitation that the Prince was prevailed upon to emerge from the marsh between the Pleisse and the Elster. An observer is doubtless more at his ease in a clock-tower than in a frail basket floating in mid-air, but steeples are not always at hand in the vicinity of battlefields and they cannot be transported at pleasure.

* * *

The first result of this treatise should be to waken the attention of men who have the mission of influencing the destinies of armies, that is to say, of governments and generals. The second, will be, perhaps, the doubling of the material and personnel of the artillery and the adoption of all improvements capable of augmenting the destructive effect. As artillerists will be among the first victims, it will be very necessary to instruct in the infantry men chosen to serve in the ranks of the artillery. Finally it will be necessary to seek means of neutralizing the effects of this carnage; the first seems to be the modification of the armament and the equipment of troops, then the adoption of new tactics which will yield results as promptly as possible.

This task will be for the rising generation, when we shall have tested by experience all the inventions with which we are occupied in the schools of artillery. Happy will be those who, in the first encounters, shall have plenty of shrapnel howitzers, many guns charged at the breech and firing thirty shots a minute; many pieces ricocheting at the height of a man and never failing their mark; finally the most improved rockets—without counting even the famous steam guns of Perkins, reserved to the defence of ramparts but which (if the written statement of Lord Wellington is to be believed) will yet be able to make cruel ravages. What a beautiful text for preaching universal peace and the exclusive reign of railroads!

KARL VON CLAUSEWITZ:

FROM

On War

Karl von Clausewitz was born in Germany in 1780 and joined the Prussian army in 1792. He studied military science in Berlin under Scharnhorst and took part in the disastrous campaign of Jena in 1806 as aide to Prince August. On release from captivity he became one of the leaders of Prussian army reform, entered Russian service and played an active part in the 1812 campaign and in the negotiations which led to Prussia's defection to the Allies. He returned to Prussian service and was chief of staff of an army corps during the Waterloo campaign.

After the peace he was appointed administrative head of the Prussian War College in 1818 with the rank of general and devoted his time to historical studies and to writing his major work, On War. *He died of cholera in 1831 and his works were published by his widow.*

His writings were found in sealed packets with a note: "Should the work be interrupted by my death, then what is found can only be called a mass of conceptions not brought into form . . . open to endless misconceptions." His writings influenced Lenin and Foch, among others.

Let us accompany the novice to the battlefield. As we approach, the thunder of the cannon becoming plainer and plainer is soon followed by the howling of shot, which attracts the attention of the inexperienced. Balls begin to strike the ground close to us, before and behind. We hasten to the hill where stands the General and his numerous Staff. Here the close striking of the cannon balls and the bursting of shells is so frequent that the seriousness of life makes itself visible through the youthful picture of imagination. Suddenly someone known to us falls—a shell strikes amongst the crowd and causes some mild panic: we begin to feel that we are no longer perfectly at ease and collected; even the bravest is at least to some degree confused.

Now, a step farther into the battle which is raging before us like a scene in the theatre, we get to the nearest General of Division; here ball follows ball, and the noise of our own guns increases the confusion. From the General of Division to the Brigadier. He, a man of acknowledged bravery, keeps carefully behind a rising ground, a house or tree—a sure sign of increasing danger. Grape rattles on the roofs of the houses and in the fields; cannon balls howl over us, and plough the air in all directions, and soon there is a frequent whistling of musket balls. A step farther towards the troops, to that sturdy infantry which for hours has maintained its firmness under this heavy fire; here the air is filled with the hissing of balls which announce their proximity by a short sharp noise as they pass within an inch of the ear, the head or the breast.

To add to all this, compassion strikes the beating heart with pity at the sight of the maimed and fallen. The young soldier cannot reach any of these different strata of danger without feeling that the light of reason does not move here in the same medium, that it is not refracted in the same manner as in speculative contemplation. Indeed, he must be a very extraordinary man who, under these impressions for the first time, does not lose the power of making any instantaneous decisions. It is true that habit soon blunts such impressions; in half an hour we begin to be more or less indifferent to all that is going on around us; but an ordinary character never attains to complete coolness and the natural elasticity of mind; and so we perceive that here again ordinary qualities will not suffice—a thing which gains truth the wider the sphere of activity which is to be filled.

Enthusiastic, stoical, natural bravery, great ambition, or also long familiarity with danger—much of all this there must be if all the effects produced in this resistant medium are not to fall short of that which may appear, to the student, only the ordinary standard.

* * *

Formerly by the term "Art of War" or "Science of War" nothing was understood but the totality of those branches of knowledge and those appliances of skill occupied with material things. . . .

In the art of sieges we first perceive a certain degree of guidance of the combat, something of the action of the intellectual faculties upon the material forces placed under their control, but generally only so far that it very soon embodied itself again in new material forms, such as approaches, trenches, counter-approaches, batteries, etc. . . .

Afterwards tactics attempted to give to the mechanism of its joints the character of a general disposition, built upon the peculiar properties of the instrument, which character leads indeed to the battlefield, but instead of leading to the free activity of the mind, leads to an Army made like an automaton by its rigid formations and orders of battle, which, movable only by the words of command, is intended to unwind its activities like a piece of clock-work.

As contemplation on War continually increased, and its history every day assumed more of a critical character, the urgent want appeared of the support of fixed maxims and rules, in order that in the controversies naturally arising about military events the war of opinions might be brought to some one point. . . . There arose, therefore, an endeavour to establish maxims, rules, and even systems for the conduct of War. . . .

All these attempts at theory are only to be considered in their analytical part as progress in the province of truth but in their synthetical part, in their precepts and rules, they are quite unserviceable.

They strive after determinate quantities, whilst in War all is undetermined, and the calculation has always to be made with varying quantities.

They direct the attention only upon material forces, while the whole action is penetrated throughout by intelligent forces and their effects.

They only pay regard to activity on one side, whilst War is a constant state of reciprocal action, the effects of which are mutual.

All that was not attainable by such miserable philosophy, the offspring of partial views, lay outside the precincts of science and was the field of genius, which *raises itself above rules.* . . .

Every theory becomes infinitely more difficult from the moment it touches on the province of moral quantities.

* * *

A swift and vigorous assumption of the offensive—the flashing sword of vengeance—is the most brilliant point in the defensive; he who does not at once think of it at the right moment, or rather he who does not from the first include this transition in his idea of the defensive will never understand the superiority of the defensive as a form of War.

* * *

If we reflect upon the commencement of War philosophically the conception of War does not originate properly with the offensive, as that form has for its absolute object, not so much *fighting* as the *taking possession of something.* The idea of war arises first by the *defensive,* for that form has the battle for its direct object, as warding off and fighting plainly are one and the same. The warding off is directed entirely against the attack; therefore supposes it, necessarily; but the attack is not directed against the warding off; it is directed upon something else—the *taking possession;* consequently does not presuppose the warding off. It lies, therefore, in the nature of things, that the party who first brings the element of War into action, the party from whose point of view two opposite parties are first conceived, also established the first laws of War, and that party is the *defender.* We are not speaking of any individual cases; we are only dealing with a general, an abstract case, which theory imagines in order to determine the course it is to take.

* * *

If we cast a glance at military history in general, we find so much the opposite of an incessant advance towards the aim, that *standing still,* and *doing nothing* is quite plainly the *normal condition* of an Army in the midst of War, *acting* is the *exception.* This must also raise a doubt as to the correctness of our conception. But . . . in the campaigns of Napoleon, the conduct of War attained to that unlimited degree of energy which we have represented as the natural law of the element. This degree is therefore possible, and if it is possible then it is necessary.

* * *

War belongs not to the province of Arts and Sciences, but to the province of social life. It is a conflict of great interests, which is settled by bloodshed, and only in that is it different from others. It would be better, instead of comparing it with any Art, to liken it to business competition, which is also a conflict of human interests and activities; and it is still more like State policy, which, again, on its part, may be looked upon as a kind of business competition on a great scale. Besides, State policy is the womb in which War is developed, in which its outlines lie hidden in a rudimentary state, like the qualities of living creatures in their germs.

If war belongs to policy, it will naturally take its character from thence. If policy is grand and powerful, so also will be the War, and this may be carried to the point at which War attains its *absolute form*.

It is true the political element does not sink deep into the details of War. Sentries are not planted, patrols do not make their rounds from political considerations; but small as is its influence in this respect, it is great in the formation of a plan for a whole War, of a campaign, and often even for a battle. . . .

In one word, the Art of War in its highest point of view is policy, but no doubt, a policy which fights battles instead of writing notes.

According to this view, to leave a great military enterprise, or the plan for one, to a *purely military* judgment and decision is a distinction which cannot be allowed, and is even prejudicial; indeed, it is an irrational proceeding to consult professional soldiers on the plan of War, that they may give a *purely military* opinion upon what the Cabinet ought to do . . . the leading outlines of a war are always determined by the Cabinet, that is . . . by a political, not a military, organ. . . .

Therefore, once more: War is an instrument of policy; it must necessarily bear its character; it must measure with its scale; the conduct of War, in its great features, is therefore policy itself, which takes up the sword in place of the pen, but does not on that account cease to think according to its own laws.

* * *

The best strategy is *always to be very strong,* first generally then at the decisive point. Therefore, apart from the energy which creates the Army, a work which is not always done by the General, there is no more imperative and no simpler law of Strategy than to *keep the forces concentrated*. No portion is to be separated from the main body unless called away by some urgent necessity. On this maxim we stand firm, and look upon it as a guide to be depended upon.

* * *

The only means of destroying the enemy's armed force is by combat, but this may be done in two ways, (1) directly, (2) indirectly, through a combination of combats. If therefore the battle is the chief means, still it is not the only means. The capture of a fortress or of a portion of territory is in itself really a destruction of the enemy's force and it may also lead to a still greater destruction and therefore, also, be an indirect means.

These means are generally estimated at more than they are worth—they have seldom the value of a battle; besides which it is always to be feared that the disadvantageous position to which they lead will be overlooked; they are seductive through the low price which they cost.

We must always consider means of this description as small investments, from which only small profits are to be expected; as means suited only to very limited State relations and weak motives. Then they are certainly better than battles without a purpose—than victories the results of which cannot be realised to the full.

STENDHAL:

FROM

The Charterhouse of Parma

Marie-Henri Beyle, who took the pseudonym of Stendhal, was born in France in 1783 and after abandoning college studies, obtained a commission in the army. After taking part in Napoleon's Italian campaign, he resigned and started to write plays, became an unsuccessful grocer, and rejoined the army on the quartermaster's staff. As such he took part in the Russian campaign. After the war he lived in Italy until he was expelled by the Austrians for suspected espionage. He settled in France in poor circumstances until he was appointed French consul in Italy after the 1830 Revolution. He died in 1842 soon after at last achieving fame with The Charterhouse of Parma, *in addition to many other works.*

The young Italian hero, in search of glory, is caught up in the battle of Waterloo, amid the confusion. . . .

We must admit that our hero was very little of a hero at that moment. However, fear came to him only as a secondary consideration; he was principally shocked by the noise, which hurt his ears. The escort broke into a gallop; they crossed a large batch of tilled land which lay beyond the canal. And this field was strewn with dead.

"Red-coats! red-coats!" the hussars of the escort exclaimed joyfully, and at first Fabrizio did not understand; then he noticed that as a matter of fact almost all these bodies wore red uniforms. One detail made him shudder with horror; he observed that many of these unfortunate red-coats were still alive; they were calling out, evidently asking for help, and no one stopped to give it them. Our hero, being most humane, took every possible care that his horse should not tread upon any of the red-coats. The escort halted; Fabrizio, who was not paying sufficient attention to his military duty, galloped on, his eyes fixed on a wounded wretch in front of him.

"Will you halt, you young fool!" the serjeant shouted after him. Fabrizio discovered that he was twenty paces on the generals' right front, and precisely in the direction in which they were gazing through their glasses. As he came back to take his place behind the other hussars, who had halted a few paces in rear of them, he noticed the biggest of these generals who was speaking to his neighbour, a general also, in a tone of authority and almost of reprimand; he was swearing. Fabrizio could not contain his curiosity; and, in spite of the warning not to speak, given him by his friend the gaoler's wife, he composed a short sentence in good French, quite correct, and said to his neighbour:

"Who is that general who is *chewing up* the one next to him?"

"Gad, it's the Marshal!"

"What Marshal?"

"Marshal Ney, you fool! I say, where have you been serving?"

Fabrizio, although highly susceptible, had no thought of resenting this insult; he was studying, lost in childish admiration, the famous Prince de la Moskowa, the "Bravest of the Brave."

Suddenly they all moved off at full gallop. A few minutes later Fabrizio saw, twenty paces ahead of him, a ploughed field the surface of which was moving in a singular fashion. The furrows were full of water and the soil, very damp, which formed the ridges between these furrows kept flying off in little black lumps three or four feet into the air. Fabrizio noticed as he passed this curious effect; then his thoughts turned to dreaming of the Marshal and his glory. He heard a sharp cry close to him; two hussars fell struck by shot; and, when he looked back at them, they were already twenty paces behind the escort. What seemed to him horrible was a horse streaming with blood that was struggling on the ploughed land, its hooves caught in its own entrails; it was trying to follow the others: its blood ran down into the mire.

"Ah! So I am under fire at last!" he said to himself. "I have seen shots fired!" he repeated with a sense of satisfaction. "Now I am a real soldier." At that moment, the escort began to go hell for leather, and our hero realised that it was shot from the guns that was making the earth fly up all round him. He looked vainly in the direction from which the balls were coming, he saw the white smoke of the battery at an enormous distance, and, in the thick of the steady and continuous rumble produced by the artillery fire, he seemed to hear shots discharged much closer at hand: he could not understand in the least what was happening.

At that moment, the generals and their escort dropped into a little road filled with water which ran five feet below the level of the fields.

The Marshal halted and looked again through his glasses. Fabrizio, this time, could examine him at his leisure. He found him to be very fair, with a big red face. "We don't have any faces like that in Italy," he said to himself. "With my pale cheeks and chestnut hair, I shall never look like that," he added despondently. To him these words implied: "I shall never be a hero." He looked at the hussars; with a solitary exception, all of them had yellow moustaches. If Fabrizio was studying the hussars of the escort, they were all studying him as well. Their stare made him blush, and, to get rid of his embarrassment, he turned his head towards the enemy. They consisted of widely extended lines of men in red, but, what greatly surprised him, these men seemed to be quite minute. Their long files, which were regiments or divisions, appeared no taller than hedges. A line of red cavalry were trotting in the direction of the sunken road along which the Marshal and his escort had begun to move at a walk, splashing through the mud. The smoke made it impossible to distinguish anything in the direction in which they were advancing; now and then one saw men moving at a gallop against this background of white smoke.

Suddenly, from the direction of the enemy, Fabrizio saw four men approaching hell for leather. "Ah! We are attacked," he said to himself; then he saw two of these men speak to the Marshal. One of the generals on the latter's staff set off at a gallop towards the enemy, followed by two hussars of the escort and by the four men who had just come up. After a little canal which they all crossed, Fabrizio found himself riding beside a serjeant who seemed a good-natured fellow. "I must speak to this one," he said to himself, "then perhaps they'll stop staring at me." He thought for a long time.

"Sir, this is the first time that I have been present at a battle," he said at length to the serjeant. "But is this a real battle?"

"Something like . . ."

The escort moved on again and made for some divisions of infantry. Fabrizio felt quite drunk; he had taken too much brandy, he was rolling slightly in his saddle: he remembered most opportunely a favourite saying of his mother's coachman: "When you've been lifting your elbow, look straight between your horse's ears, and do what the man next you does." The Marshal stopped for some time beside a number of cavalry units which he ordered to charge; but for an hour or two our hero was barely conscious of what was going on round about him. He was feeling extremely tired, and when his horse galloped he fell back on the saddle like a lump of lead.

Suddenly the serjeant called out to his men: "Don't you see the Emperor, curse you!" Whereupon the escort shouted: *"Vive l'Empereur!"* at the top of their voices. It may be imagined that our hero stared till his eyes started out of his head, but all he saw was some generals galloping, also followed by an escort. The long floating plumes of horsehair which the dragoons of the bodyguard wore on their helmets prevented him from distinguishing their faces. "So I have missed seeing the Emperor on a field of battle, all because of those cursed glasses of brandy!" This reflexion brought him back to his senses.

They went down into a road filled with water, the horses wished to drink.

"So that was the Emperor who went past then?" he asked the man next to him.

"Why, surely, the one with no braid on his coat. How is it you didn't see him?" his comrade answered kindly. Fabrizio felt a strong desire to gallop after the Emperor's escort and embody himself in it. What a joy to go really to war in the train of that hero! It was for that that he had come to France. "I am quite at liberty to do it," he said to himself, "for after all I have no other reason for being where I am but the will of my horse, which started galloping after these generals."

What made Fabrizio decide to stay where he was was that the hussars, his new comrades, seemed so friendly towards him; he began to imagine himself the intimate friend of all the troopers with whom he had been galloping for the last few hours. He saw arise between them and himself that noble friendship of the heroes of Tasso and Ariosto. If he were to attach himself to the Emperor's escort, there would be fresh acquaintances to be made, perhaps they would look at him askance, for these other horsemen were dragoons, and he was wearing the hussar uniform like all the rest that were following the Marshal. The way in which they now looked at him set our hero on a pinnacle of happiness; he would have done anything in the world for his comrades; his mind and soul were in the clouds.

—C. K. SCOTT-MONCRIEFF
(translator)

V.
THE LATER NINETEENTH CENTURY

For I dipt into the future, far as human eye could see,
Saw the Vision of the world, and all the wonder that would be;
Pilots of the purple twilight, dropping down with costly bales;
Heard the heavens fill with shouting, and there rain'd a ghastly dew
From the nations' airy navies grappling in the central blue.

—TENNYSON
"Locksley Hall"

HELMUTH VON MOLTKE:

FROM

Instructions for the Commanders of Large Formations, 1869

"The consecutive achievements of a war are not premeditated but spontaneous and guided by military instinct."

Count Helmuth von Moltke (Senior) was born in 1800 and educated in Denmark where he joined the army. In 1821 he transferred to the Prussian army as a lieutenant and for a time supported himself largely by literary work. He wrote a novel and translated Gibbon's Decline and Fall of the Roman Empire. *He entered the Turkish service and took part in the 1839 campaign against the Egyptian forces in Syria. He returned to Germany and the Prussian general staff in the same year and served as aide-de-camp to one of the Hohenzollerns in Rome, and then to the future emperor Frederick III.*

In 1857 he was appointed chief of the General Staff despite his lack of command experience and embarked, in cooperation with Bismarck and Von Roon, the war minister, on the transformation of the Prussian military system—and the creation of the German Empire. He played a significant part in the 1866 defeat of Austria and the 1870–1871 defeat of France, at the end of which he was created a field marshal. He resigned in 1888, after the accession of William II and died in 1891.

The offensive spirit kindles the spirit; but experience has proved that this state of exaltation is transformed into an opposite condition no less acute if one suffers heavy losses. . . .

It is absolutely beyond all doubt that the man who shoots without stirring has the advantage of him who fires while advancing; that the one finds protection in the ground, whereas in it the other finds obstacles; and that if to the most spirited dash one opposes a quiet steadiness, it is fire effect, nowadays so powerful, which will determine the issue. If it is possible for us to occupy such a position that the enemy for some political or military reason, or perhaps merely for national "amour propre" will decide to attack it, it seems perfectly reasonable to utilise the advantages of the defensive at first before assuming the offensive.

In the face of that craving to push on which inspires our officers and men, it is necessary that commanders should hold them in rather than urge them forward. . . .

ABRAHAM LINCOLN:

FROM

Letter to General Hooker, 1863

*Abraham Lincoln was born on a Kentucky farm in 1809.
After many years of poverty and self-education he became
a lawyer and, in 1847, a congressman. Joining the New
Republican Party and reentering politics after a five-year
interval, he was elected to the Senate in 1858 and the presi-
dency, by a narrow majority, in 1860. Shortly after the conclusion of the Civil War and his reelection for a second
term, he was assassinated in 1865.*

I have placed you at the head of the Army of the Potomac. Of
course I have done this upon what appears to me sufficient reason,
and yet I think it best for you to know that there are some things in
regard to which I am not quite satisfied with you.

I believe you to be a brave and skillful soldier, which, of course,
I like. I also believe you do not mix politics with your profession, in
which you are right. You have confidence in yourself, which is a valuable, if not an indispensable quality. You are ambitious, which,
within reasonable bounds does good rather than harm; I think that
during General Burnside's command of the Army you have taken
counsel of your ambition and thwarted him as much as you could, in
which you did a great wrong to the country and to a most meritorious
and honorable brother officer. I have heard, in such a way as to believe it, of your recently saying that both the Army and the Government needed a dictator. Of course it was not for this, but in spite of
it, that I have given you the command. Only those generals who gain
successes can set up as dictators. What I now ask of you is military
success, and I will risk the dictatorship.

The Government will support you to the utmost of its ability, which is neither more nor less than it has done and will do for all commanders. I much fear that the spirit which you have decided to infuse into the Army of criticizing their commander and withholding confidence from him will now turn upon you. I shall assist you as far as I can to put it down. Neither you nor Napoleon, if he were alive again, could get any good out of an Army while such a spirit prevails in it; and now beware of rashness. Beware of rashness, but with energy and sleepless vigilance, go forward and give us victories.

HERMAN MELVILLE:

FROM

White-Jacket

*Herman Melville was born in New York in 1819 and ran
away to sea when he was nineteen after a brief period of
school teaching. After experience in whalers and other ships
in the South Seas, he joined the frigate* United States, *in
1843, as an ordinary seaman and was discharged the following year. He devoted himself to writing, but poverty
forced him to take minor posts in the customs service
and he died in obscurity in New York in 1891.*

Published in 1850, White-Jacket *was a young man's protest
against the abuses which he had recently experienced on the
lower deck—yet concerned with the deeper meaning of life.*

While lying in the harbor of Callao, in Peru, certain rumors had
come to us touching a war with England, growing out of the long-
vexed north-eastern boundary question. In Rio these rumors increased; and the probability of hostilities induced our commodore to
authorize proceedings that closely brought home to every man on
board the *Neversink* his liability at any time to be killed at his gun.

Among other things a number of men were detailed to pass up
the rusty cannon-balls from the shot-lockers in the hold, and scrape
them clean for service. The commodore was a very neat gentleman
and would not fire a dirty shot into his foe.

It was an interesting occasion for a tranquil observer; nor was it
altogether neglected. Not to recite the precise remarks made by the
seamen while pitching the shot up the hatchway from hand to hand,
like schoolboys playing ball ashore, it will be enough to say that,

from the general drift of their discourse—jocular as it was—it was manifest that, almost to a man, they abhorred the idea of going into action.

And why should they desire war? Would their wages be raised? Not a cent. The prize-money, though, ought to have been an inducement. But of all the "rewards of virtue" prize-money is the most uncertain; and this the man-of-war's man knows. What, then, has he to expect from war? What but harder work, and harder usage than in peace; a wooden leg or arm; mortal wounds and death? Enough, however, that by far the majority of the common sailors of the *Neversink* were plainly concerned at the prospect of war, and were plainly averse to it.

But with the officers of the quarter-deck it was just the reverse. None of them, to be sure, in my hearing at least, verbally expressed their gratification; but it was unavoidably betrayed by the increased cheerfulness of their demeanor toward each other, their frequent fraternal conferences, and their unwonted animation for several days in issuing their orders. The voice of Mad Jack—always a belfry to hear—now resounded like that famous bell of England, Great Tom of Oxford. As for Selvagee, he wore his sword with a jaunty air, and his servant daily polished the blade.

But why this contrast between the forecastle and the quarter-deck, between the man-of-war's man and his officer? Because, though war would equally jeopardize the lives of both, while it held out to the sailor no promise of promotion, and what is called glory, these things fired the breast of his officers.

It is no pleasing task, nor a thankful one, to dive into the souls of some men; but there are occasions when to bring up the mud from the bottom reveals to us on what soundings we are, and on what coast we adjoin.

How were these officers to gain glory? How but by a distinguished slaughtering of their fellow-men? How were they to be promoted? How but over the buried heads of killed comrades and messmates?

This hostile contrast between the feelings with which the common seamen and the officers of the *Neversink* looked forward to this more than possible war, is one of many instances that might be quoted to show the antagonism in which they dwell. But can men, whose interests are diverse, ever hope to live together in a harmony uncoerced? . . .

Being an establishment much more extensive than the American Navy the English armed marine furnishes a yet more striking example of this thing, especially as the existence of war produces so vast an augmentation of her naval force compared with what it is in time of peace. It is well known what joy the news of Bonaparte's sudden return from Elba created among crowds of British naval officers, who had previously been expecting to be sent ashore on half-pay. Thus, when all the world wailed, these officers found occasion for thanksgiving. I urge it not against them as men—their feelings belonged to their profession. Had they not been naval officers, they had not been rejoicers in the midst of despair.

When shall the time come, how much longer will God postpone it, when the clouds, which at times gather over the horizons of nations, shall not be hailed by any class of humanity, and invoked to burst as a bomb? Standing navies, as well as standing armies, serve to keep alive the spirit of war even in the meek heart of peace. In its very embers and smoulderings they nourish that fatal fire, and half-pay officers, as the priests of Mars, yet guard the temple, though no God be there.

WILLIAM TECUMSEH SHERMAN:

FROM

Letter to Major R. M. Sawyer, 1864

William Tecumseh Sherman was born in 1820 in Ohio and graduated from West Point in 1840. He was on active service against the Indians and engaged in administration in California, but resigned from the army in 1853 and went into Tennessee and then was appointed commander of the military superintendent of a new military academy in Louisiana and the following year entered the federal army as infantry colonel, taking part in the defeat of Bull Run.

His behavior in a succession of appointments during the early years of the Civil War led to press reports of his insanity, but he was promoted to major general after the battle of Shiloh, and commanded Grant's right in the final Vicksburg campaign. He succeeded Grant in command of the army of division of the Mississippi. As such he was responsible for the taking of Atlanta, and in October 1864 he set out on his "March to the Sea," reaching Savannah at the end of the year. After further decisive operations in the Carolinas he received Johnston's surrender and the Civil War was virtually ended.

He succeeded Grant as commanding general of the army in 1869, after pacifying service in the West, and established the military training center at Fort Leavenworth, Kansas. He retired in 1883 and died in 1891.

HEADQUARTERS DEPT. OF THE TENN.,
VICKSBURG, JAN. 31, 1864

In my former letters I have answered all your questions save
one, and that relates to the treatment of inhabitants known or suspected to be hostile or "Secesh." This is in truth the most difficult
business of our army as it advances and occupies the Southern country. It is almost impossible to lay down rules, and I invariably leave
the whole subject to the local commanders, but am willing to give
them the benefit of my acquired knowledge and experience. In
Europe, whence we derive our principles of war, wars are between
kings or rulers through hired armies, and not between peoples. These
remain, as it were, neutral, and sell their produce to whatever army is
in possession.

Napoleon when at war with Prussia, Austria, and Russia
bought forage and provisions of the inhabitants, and consequently
had an interest to protect the farms and factories which ministered to
his wants. In like manner the Allied Armies in France could buy of
the French habitants whatever they needed, the produce of the soil or
manufactures of the country. Therefore, the general rule was and is
that war is confined to the armies engaged, and should not visit the
houses of families or private interests. But in other examples a
different rule obtained the sanction of historical authority. I will only
instance one, where in the siege of William and Mary the English
army occupied Ireland, then in a state of revolt. The inhabitants were
actually driven into foreign lands, and were dispossessed of their
property and a new population introduced.

To this day a large part of the north of Ireland is held by the de-
scendants of the Scotch emigrants sent there by William's order and
an act of Parliament. The war which now prevails in our land is essentially a war of races. The Southern people entered into a clear
compact of government with us of the North, but still maintained
through state organizations a species of separate existence, with separate interests, history, and prejudices. These latter became stronger
and stronger, till at last they have led to war and have developed
fruits of the bitterest kind.

We of the North are beyond all question right in our cause, but
we are not bound to ignore the fact that the people of the South have
prejudices which form a part of their nature, and which they cannot
throw off without an effort of reason or the slower process of natural

change. The question then arises, Should we treat as absolute enemies all in the South who differ from us in opinion or prejudice, kill
or banish them, or should we give them time to think and gradually
change their conduct so as to conform to the new order of things
which is slowly and gradually creeping into their country?

When men take up arms to resist a rightful authority, we are
compelled to use like force, because all reason and argument cease
when arms are resorted to. When the provisions, forage, horses,
mules, wagons, etc., are used by our enemy, it is clearly our duty and
right to take them also, because otherwise they might be used against
us. In like manner all houses left vacant by an inimical people are
clearly our right, and as such are needed as storehouses, hospitals,
and quarters.

But the question arises as to dwellings used by women, children,
and non-combatants. So long as non-combatants remain in their
houses and keep to their accustomed peaceful business, their opinions
and prejudices can in no wise influence the war, and therefore should
not be noticed; but if any one comes out into the public streets and
creates disorder, he or she should be punished, restrained, or
banished to the rear or front, as the officer in command adjudges. If
the people, or any of them, keep up a correspondence with parties in
hostility, they are spies, and can be punished according to law with
death or minor punishment.

These are well-established principles of war, and the people of
the South having appealed to *war,* are barred from appealing for protection to our constitution, which they have practically and publicly
defied. They have appealed to war, and must abide *its* rules and
laws. . . .

WILLIAM HOWARD RUSSELL:

FROM

Crimea Despatches, 1854–1855

William Howard Russell was born near Dublin in 1820 and became a reporter on the London Times *under Dulane. In 1854 he was sent to cover the activities of the British army on the outbreak of the Crimean War. Conditions were chaotic, and the army did its best to make life as uncomfortable as possible for Russell. With the laying of a cable from Varna to Bucharest, however, he was able to send the first eye-witness reports from the front by telegram.*

His reports on the Civil War, colored by his southern sympathies, were strongly attacked—and defended. "Nowhere has his liberty of speech been so furiously arraigned and his vocation so denounced as in the United States!" wrote the Times.

In later years, as editor of the Army and Navy Gazette *he continued to stimulate an informed interest in military affairs. He traveled all over the world, received by political and military leaders and was knighted in 1895. He died in 1907.*

The Charge of the Light Brigade

Sebastapol, 25 October 1854

Lord Lucan, with reluctance, gave the orders to Lord Cardigan to advance upon the guns, conceiving that his orders compelled him to do so. The noble Earl, though he did not shrink, also saw the fearful odds against him. It is a maxim of war that "cavalry never act without support," that infantry should be close at hand when cavalry carry guns, as the effect is only instantaneous, and that it is necessary to have on the flank of a line of cavalry some squadrons in column, the attack on the flank being most dangerous. The only support our light cavalry had was the reserve of heavy cavalry at a great distance behind them, the infantry and guns being far in the rear. There was no squadron in column at all and there was a plain to charge over before the enemy's guns could be reached, of a mile and a half in length.

At ten minutes past eleven, our Light Cavalry Brigade advanced. As they rushed towards the front, the Russians opened on them from the guns in the redoubt on the right with volleys of musketry and rifles. They swept proudly past, glittering in the morning sun in all the pride and splendour of war. We could scarcely believe the evidence of our senses. Surely that handful of men were not going to charge the enemy in position? Alas! it was but too true and their desperate valour knew no bounds, and far indeed was it removed from the so-called better part—discretion. They advanced in two lines, quickening their pace as they closed towards the enemy. A more fearful spectacle was never witnessed than those by who, without power to aid, beheld their heroic countrymen rushing to the arms of death.

At the distance of 1200 yards the whole line of the enemy belched forth, from thirty iron mouths, a flood of smoke and flame, through which hissed the deadly balls. Their flight was marked by instant gaps in our ranks, by dead men and horses, by steeds flying wounded or riderless across the plains. The first line was broken—it was joined by the second, they never halted or checked their speed an instant. With diminished ranks, thinned by those thirty guns, which the Russians had laid with the most deadly accuracy, with a halo of flashing steel above their heads, and with a cheer which was many a noble fellow's death-cry, they flew into the smoke of the batteries; but ere they were lost from view, the plain was strewed with their bodies and with the carcasses of horses.

They were exposed to an oblique fire from the batteries on the hills on both sides, as well as to a direct fire of musketry. Through the clouds of smoke we could see their sabres flashing as they rode up to the guns and dashed between them cutting down the gunners as they stood. To our delight we saw them returning after breaking through a column of Russian infantry and scattering them like chaff, when the flank fire of the battery on the hill swept them down, scattered and broken as they were. At the very moment when they were about to retreat a regiment of Lancers was hurled upon their flank. Colonel Shewell of the 8th Hussars, saw the danger and rode his few men straight at them, cutting his way through with fearful loss. The other regiments turned and engaged in a desperate encounter.

With courage too great almost for credence they were breaking their way through the column which enveloped them, when there took place an act of atrocity without parallel in the modern warfare of civilized nations. The Russian gunners, when the storm of cavalry passed, returned to their guns and poured murderous volleys of grape and canister on the mass of struggling men and horses. It was as much as our Heavy Cavalry Brigade could do to cover the retreat of the miserable remnant of that band of heroes as they returned to the place they had so lately quitted in all the pride of life. At thirty-five minutes past eleven not a British soldier, except the dead and the dying, was left in front of these bloody Muscovite guns.

In the Grip of Winter

Rain kept pouring down—the skies were black as ink—the wind howled over the staggering tents—the trenches were turned into dikes —in the tents the water was sometimes a foot deep—our men had neither warm nor waterproof clothing—they were out for twelve hours at a time in the trenches—they were plunged into the inevitable miseries of a winter campaign—and not a soul seemed to care for their comfort, or even for their lives. These were hard truths, which sooner or later must have come to the ear of the people of England. It was right that they should know that the wretched beggar who wandered about the streets of London in the rain led the life of a prince compared with the British soldiers who were fighting for their country and who, we were constantly assured by the home authorities, were the best appointed army in Europe.

As the year waned and winter began to close in upon us, the army suffered greatly; worn out by night work, by vigil in rain and storm, by hard labour in the trenches, they found themselves suddenly reduced to short allowance, and the excellent and ample rations they had been in the habit of receiving were cut off or miserably reduced.

* * *

Entering one of these doors I beheld such sights as few men, thank God, have ever witnessed. In a long low room supported by square pillars arched at the top and dimly lighted through shattered and unglazed windowframes, lay the wounded Russians. The wounded, did I say? No, but the dead! the rotten and festering corpses of the soldiers, who were left to die in their extreme agony, untended, uncared for, packed as close as they could be stowed, some on the floor, others on the wretched trestles and bedsteads, or pallets of straw, saturated with blood which oozed or trickled through upon the floor, mingling with the droppings of corruption. . . . Many might have been saved by ordinary care. Many lay, yet alive, with maggots crawling about or in their wounds. Many, nearly mad by the scene around them, or seeking escape from their extremest agony, had rolled away under the beds and glared out on the heart-stricken spectator. Many with legs and arms broken and twisted, the jagged splinters sticking through the raw flesh, implored aid, water, food, or pity, or deprived of speech by the approach of death or by the dreadful injuries in the head or trunk, pointed to the lethal spot. . . .

* * *

I was honoured by a good deal of abuse from some at home for telling the truth. But I could not tell lies or make things pleasant. . . . The only thing the partisans of misrule could allege was that I did not "make things pleasant" for the authorities and that, amid the filth and starvation, and deadly stagnation of the camp, I did not go about "babbling of green fields" of present abundances and of prospects of victory.

CHARLES ARDANT DU PICQ:

Battle Studies

Charles Ardant du Picq was born in 1821 and was com-
missioned in the French army in 1844. He served in the
Crimean War, being taken prisoner at Sebastopol and
later in the Syrian campaign in Africa. A colonel of an
infantry regiment, he was killed in 1870 in the Franco-
Prussian War. His Battle Studies *were published after his*
death.

The effect of an army, of one organization on another, is at the
same time material and moral. The material effect of an organization
is in its power to destroy, the moral effect is in the fear that it
inspires.

In battle, two moral forces, even more than two material forces,
are in conflict. The stronger conquers. The victor has often lost by
fire more than the vanquished. Moral effect does not come entirely
from destructive power, real and effective as it may be. It comes,
above all, from its presumed, threatening power, present in the form
of reserves threatening to renew the battle, of troops that appear on
the flank, even of a determined frontal attack.

Material effect is greater as instruments are better (weapons,
mounts, etc.) as the men know better how to use them, and as the
men are more numerous and stronger, so that in case of success they
can carry on longer.

With equal or even inferior power of destruction he will win
who has the resolution to advance, who by his formations and maneuvers can continually threaten his adversary with a new phase of
material action, who, in a word has the moral ascendancy. Moral

effect inspires fear. Fear must be changed to terror in order to van-
quish.

When confidence is placed in superiority of material means, valuable as they are against an enemy at a distance, it may be betrayed
by the actions of the enemy. If he closes with you in spite of your superiority in means of destruction, the morale of the enemy mounts
with the loss of your confidence. His morale dominates yours. You
flee. Entrenched troops give way in this manner.

At Pharsalus, Pompey and his army counted on a cavalry corps
turning and taking Caesar in the rear. In addition Pompey's army
was twice as numerous. Caesar parried the blow, and his enemy, who
saw the failure of the means of action he counted on, was demoral-
ized, beaten, lost fifteen thousand men put to the sword (while
Caesar lost only two hundred) and as many prisoners.

Even by advancing you affect the morale of the enemy. But your
object is to dominate him and make him retreat before your ascendancy, and it is certain that everything that diminishes the enemy's
morale adds to your resolution in advancing. Adopt then a formation
which permits your destructive agency, your skirmishers, to help you
throughout by their material action and to this degree diminish that
of the enemy.

Armor, in diminishing the material effect that can be suffered,
diminishes the dominating moral effect of fear. It is easy to understand how much armor adds to the moral effect of cavalry action, at
the critical moment. You feel that thanks to his armor the enemy will
succeed in getting to you.

It is to be noted that when a body actually awaits the attack of
another up to bayonet distance (something extraordinarily rare), and
the attacking troop does not falter, the first does not defend itself.
This is the massacre of ancient battle.

Against unimaginative men, who retain some coolness and consequently the faculty of reasoning in danger, moral effect will be as
material effect. The mere act of attack does not completely succeed
against such troops. (Witness battles in Spain and Waterloo.) It is
necessary to destroy them, and we are better at this than they by our
aptitude in the use of skirmishers and above all in the mad dash of
our cavalry. But the cavalry must not be treated, until it comes to so
consider itself, as a precious jewel which must be guarded against injury. There should be little of it, but it must be good.

Guibert says that shock actions are infinitely rare. Here, infinity is taken in its exact mathematical sense. Guibert reduces to nothing, by deductions from practical examples, the mathematical theory of the shock of one massed body on another. Indeed the physical impulse is nothing. The moral impulse which estimates the attacker is everything. The moral impulse lies in the perception by the enemy of the resolution that animates you.

They say that the battle of Amstetten was the only one in which a line actually waited for the shock of another line charging with the bayonets. Even then the Russians gave way before the moral and not before the physical impulse. They were already disconcerted, wavering, worried, hesitant, vacillating, when the blow fell. They waited long enough to receive bayonet thrusts, even blows with the rifle (in the back, as at Inkerman).

This done, they fled. He who calm and strong of heart awaits his enemy, has all the advantage of fire. But the moral impulse of the assailant demoralizes the assailed. He is frightened; he sets his sight no longer; he does not even aim his piece. His lines are broken without defense, unless indeed his cavalry, waiting halted, horsemen a meter apart and in two ranks, does not break first and destroy all formation.

With good troops on both sides, if an attack is not prepared, there is every reason to believe that it will fail. The attacking troops suffer more, materially, than the defenders. The latter are in better order, fresh, while the assailants are in disorder and already have suffered a loss of morale under a certain amount of punishment. The moral superiority given by the offensive movement may be more than compensated by the good order and integrity of the defenders, when the assailants have suffered losses. The slightest reaction by the defense may demoralize the attack. This is the secret of the success of the British infantry in Spain, and not their fire by rank, which was as ineffective with them as with us.

The more confidence one has in his methods of attack or defense, the more disconcerted he is to see them at some time incapable of stopping the enemy. The effect of the present improved fire arm is still limited, with the present organization and use of riflemen, to point blank ranges. It follows that bayonet charges (where bayonet thrusts never occur), otherwise attacks under fire, will have an increasing value, and that victory will be his who secures most order and determined dash. With these two qualities, too much neglected with us, with willingness, with intelligence enough to keep a firm hold on troops in immediate support, we may hope to take and to hold what we take. Do not then neglect destructive effort before using moral effect. Use skirmishers up to the last moment. Otherwise no attack can succeed. It is true it is haphazard fire, nevertheless it is effective because of its volume.

This moral effect must be a terrible thing. A body advances to meet another. The defender has only to remain calm, ready to aim, each man pitted against a man before him. The attacking body comes within deadly range. Whether or not it halts to fire, it will be a target for the other body which awaits it, calm, ready, sure of its effect. The whole first rank of the assailant falls, smashed. The remainder, little encouraged by their reception, disperse automatically or before the least indication of an advance on them. Is this what happens? Not at all! The moral effect of the assault worries the defenders. They fire in the air if at all. They disperse immediately before the assailants who are even encouraged by this fire now that it is over. It quickens them in order to avoid a second salvo.

It is said by those who fought them in Spain and at Waterloo that the British are capable of the necessary coolness. I doubt it nevertheless. After firing, they made swift attacks. If they had not, they might have fled. Anyhow the English are stolid folks, with little imagination, who try to be logical in all things. The French with their nervous irritability, their lively imagination, are incapable of such a defense.

LEO TOLSTOY:

FROM

War and Peace

*Leo Tolstoy was born at Yasnaya Polyana in 1828, and
after a brief period at the university, joined the army. As
a junior officer he fought against hill tribes in the Caucasus
and in the Crimean War—the subject of his first literary
efforts. After leaving the army in 1856, he became absorbed
in educational studies, the management of his estates, and
in writing. War and Peace, published in 1869, first established his international reputation. He repudiated the authority of the church, preached a kind of Christian communism and pacifism, and endeavored to live the life of a
peasant, despite world attention and government hostility.
In 1910 he left home by night and died a few days later at
a remote railway station.*

Pfuhl was one of those hopelessly, immutably conceited men
ready to face martyrdom for their own ideas, conceited as only a
German can be, just because it is only a German's conceit that is
based on an abstract idea-science, that is, the supposed possession of
absolute truth. The Frenchman is conceited from supposing himself
mentally and physically to be inordinately fascinating both to men
and women. An Englishman is conceited on the ground of being a
citizen of the best-constituted state in the world, and also because he
as an Englishman always knows what is the correct thing to do, and
knows that everything that he, as an Englishman does do is indis-
putably the correct thing. An Italian is conceited from being excitable
and easily forgetting himself and other people. A Russian is conceited
precisely because he knows nothing and cares to know nothing, since

he does not believe it possible to know anything fully. A conceited German is the worst of them all, and the most hardened of all, and the most repulsive of all; he imagines that he possesses the truth in a science of his own invention, which is to him absolute truth.

Pfuhl was evidently one of these men. He had a science—the theory of the oblique attack—which he had deduced from the wars of Frederick the Great; and everything he came across in more recent military history seemed to him imbecility, barbarism, crude struggles in which so many blunders were committed on both sides that those wars could not be called war at all. They had no place in his theory and could not be made a subject for science at all.

In 1806 Pfuhl had been one of those responsible for the plan of campaign that ended in Jena and Auerstadt. But in the failure of that war he did not see the slightest evidence of the weakness of his theory. On the contrary, the whole failure was to his thinking entirely due to the departures that had been made from his theory and he used to say with his characteristic gleeful sarcasm: "Didn't I always say the whole thing was going to the devil?" Pfuhl was one of those theorists who so live their theory that they lose sight of the object of the theory—its application to practice. His love for his theory led him to hate all practical considerations, and he would not hear of them. He positively rejoiced in failure for failure, being due to some departure in practice from the purity of the abstract theory, only convinced him of the correctness of his theory.

He said a few words about the present war to Prince Andrey and Tchernishev with the expression of a man who knows beforehand that everything will go wrong and is not, indeed, displeased at this being so. The uncombed wisps of hairs sticking out straight from his head behind, and the hurriedly brushed locks in front, seemed to suggest this with a peculiar eloquence.

*   *   *

Of all these men the one for whom Prince Andrey felt most sympathy was the exasperated, determined, insanely conceited Pfuhl. He was the only one of all the persons present who was unmistakably seeking nothing for himself, and harbouring no personal grudge against anybody else. He desired one thing only—the adoption of his plan, in accordance with the theory that was the fruit of years of toil. He was ludicrous; he was disagreeable, with his sarcasm, but yet he roused an involuntary feeling of respect for his boundless devotion to an idea.

Apart from this, with the single exception of Pfuhl, every speech of every person present had one common feature, which Prince Andrey had not seen at the council of war in 1805—that was, a panic dread of the genius of Napoleon, a dread which was involuntarily betrayed in every utterance now, in spite of all efforts to conceal it. Anything was assumed possible for Napoleon; he was expected from every quarter at once, and to invoke his terrible name was enough for them to condemn each other's suggestions. Pfuhl alone seemed to look on him, too, even Napoleon, as a barbarian, like every other opponent of his theory; and Pfuhl roused a feeling of pity, too, as well as respect, in Prince Andrey.

From the tone with which the courtiers addressed him, from what Paulucci had ventured to say to the Tsar and above all from a certain despairing expression in Pfuhl himself, it was clear that others knew, and he himself felt, that his downfall was at hand. And for all his conceit and his grumpy German irony, he was pitiful with his flattened locks on his forehead and his wisps of uncombed hair sticking out behind. Though he tried to conceal it under a semblance of anger and contempt, he was visibly in despair that the sole chance left him of testing his theory on a vast scale and proving its infallibility to the whole world was slipping away from him.

The debate lasted a long while, and the longer it continued the hotter it became, passing into clamour and personalities, and the less possible it was to draw any sort of general conclusion from what was uttered. Prince Andrey simply wondered what they were all saying as he listened to the confusion of different tongues, and the propositions, the plans, the shouts, and the objections. The idea which had long ago and often occurred to him during the period of his active service, that there was and could be no sort of military science, and that therefore there could not be such a thing as military genius, seemed to him now to be an absolutely obvious truth.

What theory and science can there be of which the conditions and circumstances are uncertain and can never be definitely known, in which the strength of the active forces engaged can be even less definitely measured? No one can, or possibly could, know the relative positions of our army and the enemy's in another twenty-four hours, and no one can gauge the force of this or the other detachment. Sometimes when there is no coward in front to cry, "We are cut off!"

and to run, but a brave spirited fellow leads the way, shouting "Hurrah!" a detachment of five thousand is as good as thirty thousand, as it was at Schöngraben, while at times fifty thousand will run from eight thousand, as they did at Austerlitz. How can there be a science of war in which, as in every practical matter, nothing can be definite, and everything depends on countless conditions, the influence of which becomes manifest all in a moment, and no one can know when that moment is coming?

Armfeldt declares that our army is cut off, while Paulucci maintains that we have caught the French army between two fires; Michaud asserts that the defect of the Drissa camp is having the river in its rear, while Pfuhl protests that that is what constitutes its strength; Toll proposes one plan, Armfeldt suggests another; and all are good and all are bad and the suitability of any proposition can only be seen at the moment of trial.

And why do they all talk of military genius? Is a man to be called a genius because he knows when to order biscuits to be given out, and when to march his troops to the right and when to the left? He is only called a genius because of the glamour and authority with which the military are invested, and because masses of sycophants are ready to flatter power, and to ascribe to it realities quite alien to it. The best generals I have known are, on the contrary, stupid or absent-minded men. The best of them is Bagration—Napoleon himself admitted it.

And Bonaparte himself! I remember his fatuous and limited face on the field of Austerlitz. A good general has no need of genius, nor of any great qualities; on the contrary, he is the better for the absence of the finest and highest of human qualities—love, poetry, tenderness, philosophic and inquiring doubt. He should be limited, firmly convinced that what he is doing is of great importance (or he would never have patience to go through with it), and only then will he be a gallant general. God forbid he should be humane, and should feel love and compassion, should pause to think what is right and wrong. It is perfectly comprehensible that the theory of their genius should have been elaborated long, long ago, for the simple reason that they are the representatives of power. The credit of success in battle is not by right theirs; for victory or defeat depends in reality on the soldier in the ranks who first shouts "Hurrah!" or "We are lost!" And it is only in the ranks that one can serve with perfect conviction, that one is of use!

* * *

Napoleon could not command a campaign against Russia and never did command it. He commanded one day certain papers to be written to Vienna, to Berlin and to Petersburg; next day, certain decrees and instructions to the army, the fleet, and the commissariat, and so on and so on—millions of separate commands, making up a whole series of commands, corresponding to a series of events leading the French soldiers to Russia.

Napoleon was giving commands all through his reign for an expedition to England. On no one of his undertakings did he waste so much time and so much effort, and yet not once during his reign was an attempt made to carry out his design. Yet he made an expedition against Russia, with which, according to his repeatedly expressed conviction, it was to his advantage to be in alliance; and this is due to the fact that his commands in the first case did not, and in the second did, correspond with the course of events. . . .

Our false conception that the command that precedes an event is the cause of an event is due to the fact that when the event has taken place and those few out of the thousands of commands, which happen to be consistent with the course of events, are carried out, we forget those which were not, because they could not be carried out. Apart from that, the chief source of our error arises from the fact that in the historical account a whole series of innumerable, various, and most minute events, as for instance, all that led the French soldiers to Russia, are generalised into a single event, in accordance with the result produced by that series of events: and by a corresponding generalisation a whole series of commands, too, is summed up into a single expression of will.

For causes, known or unknown to us, the French begin to chop and hack at each other. And to match the event, it is accompanied by its justification in the expressed wills of certain men, who declare it essential for the good of France, for the cause of freedom, of equality. Men cease slaughtering one another, and that event is accompanied by the justification of the necessity of centralisation of power, of resistance to Europe, and so on. Men march from west to east, killing their fellow-creatures and this event is accompanied by phrases about the glory of France, the baseness of England, and so on. History teaches us that those justifications for the event are devoid of all common-sense, that they are inconsistent with one another, as, for instance, the murder of a man as a result of the declaration of his rights, and the murder of millions in Russia for the abasement of England. But those justifications have an incontestable value in their own day.

—CONSTANCE GARNETT
(translator)

IVAN STANISLAVOVICH BLOCH:

FROM

Modern Weapons and Modern War

Ivan Stanislavovich Bloch, also known as Jean de Bloch, was born in Radom, Poland, in 1836. For most of his life he was an official of the railroad administration in then Russian Poland, but became an ardent advocate of universal peace and founded a Peace Museum in Lucerne, Switzerland. His book The Future of War in Its Technical, Economic and Political Relation *was published in 1898 in seven volumes. An abridged English translation by R. C. Long was brought out in London in 1899 under the title* Is War Now Impossible? *His* Modern Weapons and Modern War *followed in 1900 and he died in 1902. He was one of the first to hold that warfare would become so terrible and so costly that no nation could afford to undertake it.*

As we have already explained, the quick and final decision of future battles is improbable. The latest improvements in small arms and artillery, and the teaching of troops to take advantage of localities, has increased the strength of defence. The modern rifle has immense power, and its use is simple and convenient. It will be extremely difficult to overcome the resistance of infantry in sheltered positions. Driven from one position it will quickly find natural obstacles—hillocks, pits, and groups of trees—which may serve as points for fresh opposition. The zone of deadly fire is much wider than before, the battles will be more stubborn and prolonged. Of such a sudden sweeping away of an enemy in the course of a few minutes as took place at Rossbach it is absurd even to think.

The power of opposition of every military unit has increased so greatly that a division may now accept battle with a whole army corps, if only it be persuaded that reinforcements are hastening to the spot. The case already cited, of the manoeuvres in Eastern Prussia, when a single division sustained an attack from a whole army corps until reinforced, is sufficient evidence of this. The scattering of immense masses over a considerable space means that a successful attack on one point by means of the concentration of superior forces may remain local, not resulting in any general attack on the chief forces of the defence.

In former times either of the combatants quickly acknowledged that the advantage lay with the other side, and therefore refused to continue the battle. The result and the trophy of victory was the possession of the battlefield. The majority of military writers consider the attainment of such a result very questionable.

From the opinions of many military writers the conclusion is inevitable that with the increase of range and fire, and in view of the difficulties with which assault is surrounded, a decisive victory in the event of numerical equality is possible only on the failure of ammunition on one side. But in view of the number of cartridges which soldiers now carry, and the immense reserves in the ammunition carts, it seems more likely, that before all cartridges have been expended, the losses will have been so great as to make a continuation of battle impossible. To the argument that night will interrupt the battle we find an answer in the fact that, thanks to the adoption of electric illuminations, the struggle will often continue or be renewed at night.

In all armies attempts are made to inspire the soldiers with the conviction that a determined assault is enough to make an enemy retreat. Thus, in the French field instructions we find it declared that "courageous and resolutely led infantry may assault, under the very strongest fire, even well-defended earthworks and capture them." But the above considerations are enough to show the difficulty of such an undertaking.

Supposing even that the defenders begin a retreat. The moment the attacking army closes its ranks for assault partisan operations on the side of the defenders will begin. Indeed, it may be said that the present rifle, firing smokeless powder, is primarily a partisan weapon, since armed with it even a small body of troops in a sheltered position may inflict immense losses from a great distance. As the attackers approach, the thin flexible first line of the defence will retreat. It will annoy the enemy with its fire, forcing him to extend his formation, and then renew the manoeuvre at other points.

While the first line of the defenders will thus impede the assault, the main body will have opportunity to form anew and act according to circumstances. The attacking army, though convinced of victory, finding that it cannot get into touch with the rear-guard of the enemy, which alternately vanishes and reappears, now on its flanks, now in front, will lose confidence, while the defenders will take heart again.

It is obvious that, with the old powder, the smoke of which betrayed the fighting front of the enemy and even approximately indicated its strength, such manoeuvres were too dangerous to carry out. It would be a mistake to think that for the carrying on of such operations picked troops are required. The ordinary trained soldier is quite capable. Every soldier knows that two or three brigades cannot entirely stop the advance of an army. But seeing that the attackers may be so impeded that they will gain no more than four or five miles in a day, the defenders will have good cause to hope and wait for a favourable turn of affairs.

From this it may be seen how immensely smokeless powder has increased the strength of defence. It is true that in past wars we find many examples of stubborn rear-guard actions facilitating orderly retreat. But even in those cases victory was too evident and irrevocable, and this encouraged the pursuers. The vanquished tried as quickly as possible to get out of fire. Nowadays with quick-firing and long-range guns the first few miles of retreat will prove more dangerous than the defence of a position, but the chain of marksmen covering the retreat may greatly delay the course of the attack.

It was Marshal St. Cyr who declared that "a brave army consists of one-third of soldiers actually brave, one-third of those who might be brave under special circumstances, and a remaining third consisting of cowards." With the increase of culture and prosperity nervousness has also increased, and in modern, especially in Western European armies, a considerable proportion of men will be found unaccustomed to heavy physical labour and to forced marches. To this category the majority of manufacturing labourers will belong. Nervousness will be all the more noticeable since night attacks are strongly recommended by many military writers, and undoubtedly these will be made more often than in past wars. Even the expectation of a battle by night will cause alarm and give birth to nervous excitement.

This question of the influence of nervousness on losses in time of war has attracted the attention of several medical writers, and some have expressed the opinion that a considerable number of soldiers will be driven mad. The famous Prussian Minister of War, Von Roon, writing from Nikelsburg in 1866, said: "Increased work and the quantity and variety of impressions have so irritated my nerves that it seems as if fires were bursting out in my brain."

We have already referred many times to the probability of prolonged wars in the future. Against this probability only one consideration may be placed: the difficulty of provisioning immense armies and the probability of famine in those countries which in times of peace live upon imported corn. With the exception of Russia and Austria-Hungary, not a single country in Europe is in a position to feed its own population. Yet Montecuculli said: "Hunger is more terrible than iron, and want of food will destroy more armies than battles." Frederick II declared that the greatest military plans might be destroyed by want of provisions. But the army of Frederick II was a mere handful in comparison with the armies of to-day.

It is true that ancient history presents examples of immense hordes entering upon war. But these wars were generally decided by a few blows, for there existed neither rapid communications for the purpose of reinforcement, nor regular defensive lines. Modern history shows many instances of prolonged wars. But it must be remembered that the Thirty Years' and the Seven Years' wars were not uninterrupted, and that the armies engaged went into winter quarters where they were regularly provisioned, and in spring recommenced operations resulting only in partial successes, the gaining of a battle, the taking of a fortress, followed by another stoppage of operations. Thus the long wars of modern history may be regarded as a series of short campaigns.

ALFRED VON SCHLIEFFEN:

FROM

The Great Memorandum

*Alfred von Schlieffen was born in 1833, and after briefly
studying law joined the Prussian army in 1854. He became
a member of the General Staff in 1865, and thereafter he
devoted himself almost entirely to it. He became its chief
in 1891, retired in 1905—though he continued to advise it—
and died in 1913. His last words were said to be, "The
struggle is inevitable. Keep my right flank strong!"*

*It is also recorded that on one occasion, while traveling
by train, his aide attempted to make conversation by com-
menting on the beauty of the mountain valley. "An insignificant obstacle" was the only rejoinder. He spent most
of his life working at his "Plan" which was eventually put
into operation by his successor in 1914.*

*The Schlieffen Plan went through a number of revisions be-
fore its final acceptance by the German General Staff in
1905. The conception was said to have derived from
Cannae.*

*The actual distribution of forces and the details of
maneuver were left uncertain. When it was carried into
effect by the younger Moltke in 1914, it achieved only a
limited success.*

In a war against Germany, France will probably at first restrict
herself to defence, particularly as long as she cannot count on effective Russian support. With this in view she has long prepared a position which is for the greater part permanent, of which the great
fortresses of Belfort, Epinal, Toul and Verdun are the main strong-

points. This position can be adequately occupied by the large French army and presents great difficulties to the attacker.

The attack will not be directed on the great fortresses, whose conquest requires a great siege apparatus, much time and large forces, especially as encirclement is impossible and the siege can only be conducted from one side. The attacker will prefer to advance on the intervening gaps. . . .

Therefore a frontal attack on the position Belfort-Verdun offers little promise of success. An envelopment from the south would have to be proceeded by a victorious campaign against Switzerland and by the capture of the Jura forts—time-consuming enterprises during which the French would not remain idle.

Against a northern envelopment the French intend to occupy the Meuse between Verdun and Mezières, but the real resistance, it is said, is not to be offered here but behind the Aisne, roughly between St. Menehould and Rethel. An intermediate position beyond the Aire seems also to be under consideration. If the German envelopment reaches even further, it will run into a strong mountainous position whose strongpoints are the fortresses of Rheims, Laon and La Fère. . . .

One cannot have great confidence in an attack on all these strong positions. More promising than the frontal attack with an envelopment by the left wing seems to be an attack from the north-west directed on the flanks, at Mezières, Rethel, La Fère, and across the Oise on the rear of the position.

To make this possible, the Franco-Belgian frontier left of the Meuse must be taken, together with the fortified towns of Mezières, Hirson, and Maubeuge, three small barrier forts, Lille and Dunkirk; and to reach thus far the neutrality of Luxembourg, Belgium and the Netherlands must be violated.

The violation of Luxembourg neutrality will have no important consequence other than protests. The Netherlands regard England, allied to France, no less as an enemy than does Germany. It will be possible to come to an agreement with them.

Belgium will probably offer resistance. In face of the German advance north of the Meuse, her army, according to plan, will retreat to Antwerp and must be contained there; this might be effected in the north by means of a blockade of the Scheldt which would cut communications with England and the sea. For Liège and Namur, which are intended to have only a weak garrison, observation will suffice. It will be possible to take the citadel of Huy or to neutralise it.

—ANDREW and EVA WILSON
(translators)

Moltke's Comments on the Schlieffen Plan, 1911

It is important, of course, that for an advance through Belgium the right wing should be made as strong as possible. But I cannot agree that the envelopment demands the violation of Dutch neutrality in addition to Belgian. A hostile Holland at our back could have disastrous consequences for the advance of the German army to the west, particularly if England should use the violation of Belgian neutrality as a pretext for entering the war against us. A neutral Holland secures our rear, because if England declares war on us for violating Belgian neutrality, she cannot herself violate Dutch neutrality. She cannot break the very law for whose sake she goes to war.

Furthermore it will be very important to have in Holland a country whose neutrality allows us to have imports and supplies. She must be the windpipe that enables us to breathe.

EMORY UPTON:

FROM

The Military Policy of the United States

Emory Upton was born in New York in 1839 and graduated from West Point. He distinguished himself in combat with the Federal army in the Civil War, serving with the infantry, cavalry, and artillery and finishing as a brevet major general.

He devised a new system of infantry tactics which was adopted in 1867. From 1870 to 1875 he was commandant of cadets and instructor in artillery, cavalry, and infantry at West Point. He then undertook a world tour, at Sherman's request, to study foreign military systems and produced his Armies of Asia and Europe *on his return in 1875.*

Shortly after his appointment to general command in San Francisco in 1881 he shot himself at the age of forty-one. He was a lonely man, intensely religious and frustrated by the state of the American army and America.

Upton's study was unfinished when he shot himself and it remained on file for many years. "His recommendations had all the prestige of his brilliant career," said Elihu Root, the Secretary of War, in 1903. "They had the advocacy and support of the great soldier who commanded the American armies, General Sherman. . . . Yet his voice was as the voice of one crying in the wilderness."

The Lessons to be drawn from the Revolution are:
First that nearly all of the dangers which threatened the cause of

independence may be traced to the total inexperience of our statesmen in regard to military affairs, which led to vital mistakes in army
legislation. . . .

Ninth. That the draft, connected or not connected, with volun-
tary enlistments and bounties, is the only sure reliance of a government in time of war.

Tenth. That short enlistments are destructive to discipline, constantly expose an army to disaster, and inevitably prolong war with
all its attendant dangers and expenses.

Eleventh. That short enlistments at the beginning of a war tend
to disgust men with the service, and force the Government to resort
either to bounties or to the draft.

Twelfth. That regular troops, engaged for war, are the only safe
reliance of a government, and are in every point of view the best and
most economical.

Thirteenth. That when a nation at war relies upon a system of
regulars and volunteers, or regulars and militia, the men, in the absence of compulsion, or very strong inducements, will invariably enlist in the organizations most lax in discipline.

Fourteenth. That troops become reliable only in proportion as
they are disciplined; that discipline is the fruit of long training, and
cannot be attained without the existence of a good corps of officers.

Fifteenth. That the insufficiency of numbers to counterbalance a
lack of discipline should convince us that our true policy, both in
peace and war, as Washington puts it "ought to be to have a good
army rather than a large one."

In seeking to trace all the great mistakes and blunders committed during the Civil War to defects of our military system, it is important to bear in mind the respective duties and responsibilities of soldiers and statesmen. The latter are responsible for the creation and
organization of our resources, and, as in the case of the President,
may further be responsible for their management or mismanagement.
Soldiers, while they should suggest and be consulted on all details of
organization under our system, can alone be held responsible for the
control and direction of our armies in the field.

So long as historians insist upon making our commanders alone
responsible for disasters in time of war, so long will the people and
their representatives fail to recognize the importance of improving
our system.

ALFRED THAYER MAHAN:

FROM

The Influence of Sea Power Upon the French Revolution

Alfred Thayer Mahan was born at West Point in 1840 and after graduation from Annapolis served in the United States Navy, seeing active service in the Civil War and designing a "mystery ship." He became a lecturer on naval history and strategy at the Naval War College and its president in 1886. He commanded the cruiser Chicago, *served on the naval war board during the Spanish-American War, and was a delegate to the Hague Peace Conference in 1899.*

By this time he had gained a world reputation through his writings on strategy and, in Britain especially, as the celebrant of maritime supremacy. The Kaiser quoted Mahan in support of German naval expansion. He advocated the qualitative limitation of naval armaments—but believed in the inevitability of war and in Anglo-American cooperation. Theodore Roosevelt sought his help in his plans for attack on Cuba; he was an influential advocate of the annexation of Hawaii and the retention of the Philippines. A rear-admiral, he died in 1914, shortly after the outbreak of World War I.

Meanwhile that period of waiting from May, 1803, to August, 1805, when the tangled net of naval and military movements began to unravel, was a striking and wonderful pause in the world's history. On the heights above Boulogne, and along the narrow strip of beach from Etaples to Vimereux, were encamped one hundred and thirty thousand of the most brilliant soldiery of all time, the soldiers who had fought in Germany, Italy, and Egypt, soldiers who were yet to win, from Austria, Ulm and Austerlitz, and from Prussia, Auerstadt and Jena, to hold their own, though barely, at Eylau against the army of Russia, and to overthrow it also, a few months later, on the bloody field of Friedland.

Growing daily more vigorous in the bracing sea air and the hardy life laid out for them, they could on fine days, as they practiced the varied maneuvers which were to perfect the vast host in embarking and disembarking with order and rapidity, see the white cliffs fringing the only country that to the last defied their arms.

Far away, Cornwallis off Brest, Collingwood off Rochefort, Pellew off Ferrol, were battling the wild gales of the Bay of Biscay, in that tremendous and sustained vigilance which reached its utmost tension in the years preceding Trafalgar, concerning which Collingwood wrote that admirals need to be made of iron, but which was forced upon them by the unquestionable and imminent danger of the country. Farther distant still, severed apparently from all connection with the busy scene at Boulogne, Nelson before Toulon was wearing away the last two years of his glorious but suffering life, fighting the fierce north-westers of the Gulf of Lyon and questioning, questioning continually with feverish anxiety, whether Napoleon's object was Egypt again or Great Britain really.

They were dull, weary, eventless months, those months of watching and waiting of the big ships before the French arsenals. Purposeless they surely seemed to many, but they saved England. The world has never seen a more impressive demonstration of the influence of sea power upon its history. Those far distant, storm-beaten ships, upon which the Grand Army never looked, stood between it and the dominion of the world.

Holding the interior positions they did, before—and therefore between—the chief dockyards and detachments of the French navy, the latter could unite only by a concurrence of successful evasions, of which the failure of any one nullified the result. Linked together as the various British fleets were by chains of smaller vessels, chance alone could secure Bonaparte's great combination, which depended upon the covert concentration of several detachments upon a point practically within the enemy's lines. Thus, while bodily present before Brest, Rochefort, and Toulon, strategically the British squadrons lay in the Straits of Dover barring the way against the Army of Invasion.

The Straits themselves, of course, were not without their own special protection. Both they and their approaches, in the broadest sense of the term, from the Texel to the Channel Islands, were patrolled by numerous frigates and smaller vessels, from one hundred to a hundred and fifty in all. These not only watched diligently all that happened in the hostile harbors and sought to impede the movements of the flat-boats, but also kept touch with and maintained communication between the detachments of ships-of-the-line. Of the latter, five off the Texel watched the Dutch navy, while others were anchored off points of the English coast with reference to probable movements of the enemy.

Lord St. Vincent, whose ideas on naval strategy were clear and sound, though he did not use the technical terms of the art, discerned and provided against the very purpose entertained by Bonaparte, of a concentration before Boulogne by ships drawn from the Atlantic and Mediterranean. The best security, the most advantageous strategic positions, were doubtless those before the enemy's ports; and never in the history of blockades has there been excelled, if ever equaled, the close locking of Brest by Admiral Cornwallis, both winter and summer, between the outbreak of war and the battle of Trafalgar. It excited not only the admiration but the wonder of contemporaries.

COLMAR VON DER GOLTZ:

FROM

The Nation in Arms

Baron Colmar Von der Goltz was born in 1843 and joined the Prussian army in 1861. After a short period of active service in the Austrian War he became a member of the general staff. In 1871 he was appointed a professor at the military school in Potsdam and in 1878 a lecturer in military history at the military academy in Berlin. He spent twelve years in Turkey reorganizing military establishments and was promoted field marshal in 1911 and retired in 1913.

After the outbreak of war he was made governor-general of occupied Belgium and then was sent to Turkey. He was put in command of an army in the field, for the first time in his life, against the British in Mesopotamia. He said he would bring Townshend, the British commander, back to Constantinople "dead or alive." Townshend eventually surrendered at Kut and was brought back to Constantinople —and Von der Goltz accompanied him on the train, having died a few days earlier.

We frequently hear the complaint that all advances made by modern science and technical art are immediately applied to the abominable end of annihilating mankind. Instead, so it would appear, of nations rising by such advances ever higher and higher in civilization, they thus become only ruder and more brutal, since they brood on mutual destruction with ever-growing zeal. But this assertion is only seemingly correct. The nobler and grander life of a nation has become by culture, science, art and wealth, the more it has to lose by war; it will, consequently, be more careful to equip itself thoroughly for battle. . . .

As a rule, high culture and military power go hand-in-hand, as evidenced in the cases of Greece and Rome. Again, we must not advance exceptions like England, whose military system is out of all proportion to the development of the State in other respects. Protected by the sea, it has only colonial wars to wage, in which money plays the chief part. This latter is the sharpest weapon in England's hand. In addition to this, it maintains a fleet such as no other Empire can boast. But, in spite of all the advantages of its position, it will soon find itself compelled to follow the lead of the times, and to strengthen the organization of its army, or it will gradually sink both in power and influence upon the Continent.

But, as the fusion of the military system with the national and political life leads, when compared with the results, to a diminution of the sacrifices demanded, so, despite all appearances, war becomes more humane by taking advantage of the progress of civilization. The foe is conquered, not by destruction of his existence, but by the annihilation of his hopes of victory.

"Fighting to the last man," as we may add to soothe timid minds, is only a strong figure of speech expressing a determination to fight bravely. It would sound strange if an army were to vow before battle, to fight until it had lost twenty per cent; and yet this would be more, and much more, than sufficient. As a rule, a loss of half this proportion on either side is sufficient to decide the victory. The destruction of a part of the whole deters the rest from further exertions, and ends the struggle. The more startling and intense the effects of the weapons, the sooner do they produce a deterrent effect, and thus it comes to pass that battles generally are less bloody in proportion as the engines of destruction have attained greater perfection.

HANS DELBRÜCK:

FROM

History of the Art of War

"The Strategy of Annihilation has only one pole, the battle, whereas the Strategy of Exhaustion has two poles, battle and manoeuvre. . . ."

Hans Delbrück was born in 1848, attended several universities and enlisted in the Prussian army in 1867, seeing active service in the Franco-Prussian War. In 1874 he became tutor to the son of the Crown Prince and thereafter devoted himself to military study, apart from a short term as a member of the Reichstag. The application of scientific methods to military history constituted his major contribution but he did much to stimulate interest in military affairs through numerous articles. Although a German nationalist, he became critical of government policy and during World War I advocated a negotiated peace with the Western Allies. He died in 1929.

To discover in the abstract, with the benefit of hindsight and the knowledge of all the facts, the best possible source of action is not really so difficult. Great military ideas are actually extremely simple. The most celebrated manoeuvres, counted by history as the work of true genius, for instance the Prussian withdrawal from Ligny *towards* Waterloo, could be invented on the map by a regimental clerk. Greatness lies in the freedom of the intellect and spirit at moments of pressure and crisis, and in the willingness to take risks.

The military critic, therefore, won't hesitate to show us that at Borodino Napoleon acted too cautiously and with insufficient determination when he contented himself with forcing the Russians to re-treat instead of committing his last reserves and gaining complete vic-tory. The military critic tells us that at Waterloo the French army would have been not only broken up but captured if the Prussians had left Planchenoit an hour earlier. . . .

The historian attempts to evoke an understanding of the mag-nitude of the career that at Borodino reached both its climax and the beginning of its collapse. He seeks to show the gigantic individual who had conquered the world and who out of a new sense of in-adequacy for the first time set a limit to his own success. In the pres-ence of sublime truth criticism should fall silent.

In Napoleon's cautious refusal to take the last necessary step, a refusal that inevitably leads to disaster, it is not weakness that we see but human nature. Embarrassment would take the place of awe if the victory paean of Waterloo were to close with the observation that the triumph could have been greater if the Allies had avoided this or that error. We do not deny that errors were made, but they are errors only in the eyes of the military critic. To the historian they appear as facets of the conflicting forces, which he depicts as they are, not as they would like to be. . . .

In his theoretical writings Clausewitz put forward with particu-lar insistence the proposition that the purpose of war (in the nar-rower, not in the political sense) is the destruction of the enemy's military power. Consequently battle must be considered as the only decisive factor in war, the goal of all strategy. This precept is Clause-witz's real legacy to the Prussian army; it is the sum and conclusion of all his arguments. All previous strategic systems were based on the opposite view: that success in war might also be achieved through manoeuvring, occupying positions, the careful design of one's own communications and the disruption of the enemy's.

The type of warfare from which Clausewitz abstracted his law, like Lessing abstracted the laws of poetry from Homer, is Napoleonic warfare. But Frederick the Great's campaigns were in their essentials obviously based on the opposite system. Clausewitz himself repeat-edly comes to the conclusion that during the Frederickian period of warfare battle was regarded as an evil to which men subjected them-selves only when there was no help for it. . . .

If Clausewitz from the outset had proceeded as an historian he undoubtedly would soon have found that the political and military in-stitutions of the 18th century offered compelling reasons why the age should adopt the manoeuvre system, just as in the 19th century new political and social forces enabled the stronger and purer principle of combat to triumph. While Clausewitz did not recognise this as universal, but saw it only in particular cases, his analyses hardly suffered, since he did understand that the dependence of even the greatest general on the prevalent views of his generation is a factor in his decision.

—PETER PARET
(translator)

FERDINAND FOCH:

FROM

The Principles of War

Ferdinand Foch was born in 1851 and joined the French army in 1870. Most of his army service was spent in garrison duty and on the staff. He became a lecturer in military history at the Ecole de Guerre in 1895 and its director in 1908. Then in 1913 he was appointed to command an elite corps on the German frontier by a government which had been impressed by his offensive ideas and his knowledge of German thinking.

At the outbreak of the war he made a successful counterattack in Lorraine, after an initial reverse, and was appointed to the command of an army. Despite criticism of his conduct of operations he was promoted commander of the Northern Army Group the following year, after heavy Allied losses. He replaced Pétain as Chief of the General Staff in 1917, when the latter took command at the front. After the last German offensive in 1918, he was appointed supreme commander of the Allied forces, which now included the Americans, and he launched the final offensive.

His role both in the final war operations and in the peace arrangements continued to be a matter of widespread controversy. He was made a marshal of France and died in 1929.

Based on his lecture notes while he was a Staff College in-structor (and incorporating a detailed battle analysis from a German source), the Principles *reflect French "offensive" thought.*

We cannot draw our inspiration indifferently from Turenne, Condé, Prince Eugene, Villars, or Frederick the Great, even less from the tottering theories and degenerate forms of the last century. The best of these doctrines answered a situation and needs which are no longer ours.

Our models, and the facts on which we will base a theory, we must seek in certain definite pages of history, namely from that period of the French Revolution when the whole nation was arming itself for the defence of its dearest interests: Independence, Liberty; from that period of the Empire, when the army born of that violent crisis was taken in hand and led by the greatest military genius that ever was, and thus gave rise to the matchless masterpieces of our art.

No strategy can henceforth prevail over that which aims at ensuring tactical results, victory by fighting. A strategy paving the way to tactical decisions alone: this is the end we come to in following a study which has produced so many learned theories. Here, as everywhere else, as in politics, the entrance upon the stage of human masses and passions necessarily leads to simplification.

—HILAIRE BELLOC
(translator)

GEORGE FRANCIS HENDERSON:

FROM

Stonewall Jackson

George Francis Henderson was born in 1852 and after Oxford and Sandhurst joined the army in 1878. After active service in India and Egypt he was posted to North America and developed a particular interest in the Civil War. In 1889 he was appointed instructor in Tactics, Military Administration and Law, at Sandhurst. In 1892 he was transferred to the Staff College as professor of Military Art and History. At the outbreak of the war in South Africa, he was appointed Director of Intelligence to the British Commander in Chief. His health deteriorated and he died in 1903. As a teacher he exercised a considerable influence on many officers who were to reach high command.

The two-volume study of Stonewall Jackson focused the attention of a generation of soldiers, especially in Britain, on the detailed lessons of the American Civil War.

It is true that Jackson's force was very small. But the manifestation of military genius is not affected by numbers. The handling of masses is a mechanical art, of which knowledge and experience are the key, but it is the manner in which the grand principles of war are applied which make the great leader, and these principles may be applied as resolutely and effectively with 10,000 men as with 100,000.

"In meditation" says Bacon "all dangers should be seen; in execution, none, unless very formidable." It was on this precept that Jackson acted. Not a single one of his manoeuvres but was based on a close and judicial survey of the situation. Every risk was weighed.

Nothing was left to chance. The character of his opponent, the morale of the hostile troops, the nature of the ground, and the manner in which physical features could be turned to account were all matters of the most careful consideration.

It is little wonder that it should have been said by his soldiers that "he knew every hole and corner of the Valley as if he had made it himself". . . .

The first principle of war is to concentrate superior force at the decisive point, that is, upon the field of battle. But it is exceedingly seldom that by standing still, and leaving the initiative to the enemy, that this principle can be observed, for a numerically inferior force, if it once permits its enemy to concentrate, can hardly hope for success. True generalship is, therefore "to make up in activity for lack of strength"; to strike the enemy in detail and overthrow his columns in succession. And the highest art of all is to compel him to disperse his army, and then to concentrate superior force against each fraction in turn.

JULIAN CORBETT:

FROM

Some Principles of Maritime Strategy

Julian Corbett was born in 1854 and, after graduating from Cambridge, practiced law for a time and published several novels. He became a lecturer in history at the Naval War College and established a reputation through his lectures and writing as a leading naval historian and a successor to Mahan as an upholder of maritime supremacy. In 1914 he produced, in collaboration with Admiral Fisher, "the Baltic Scheme"—a plan to sow the North Sea with mines and bring about a Russian amphibious landing on the north coast of Germany. The Government suppressed the paper and rejected the scheme. Corbett was knighted in 1917 and died in 1922.

Although he was proven wrong about submarines and convoys, Corbett provided an influential followup to Mahan's exposition of sea power.

The paramount concern, then, of maritime strategy is the determining the mutual relations of your army and navy in a plan of war. The theory which now holds the field is that war in a fundamental sense is continuation of policy by other means. . . .

The real secret of Wellington's success—apart from his own genius—was that in perfect conditions he was applying the limited form to an unlimited war. Our object was unlimited. It was nothing less than the overthrow of Napoleon. Complete success at sea had failed to do it, but that success had given us the power of applying the limited form, which was the most decisive form of offence within our means. . . .

Concentration should be so arranged that any two parts may freely cohere, and that all parts may quickly condense into a mass at any point in the area of concentration. The object of holding back from forming the mass is to deny the enemy knowledge of our actual distribution or its intention at any given moment, and at the same time to ensure that it will be adjusted to meet any dangerous movement that is open to him. Further than this our aim should be not merely to prevent any part being overpowered by a superior force but to regard every detached squadron as a trap to lure the enemy to destruction. The ideal concentration, in short, is an appearance of weakness that covers a reality of strength.

* * *

The unproved value of submarines only deepens the mist which overhangs the next naval war. From a strategical point of view we can say no more than that we have to count with a new factor which gives a new possibility to minor counter-attacks. It is a possibility which, on the whole, tells in favour of naval defence, a new card which skilfully played in combination with defensive fleet operations, may lend fresh importance to the "Fleet in being". . . .

Upon the great routes the power of attack has been reduced and the means of evasion has increased to such an extent as to demand entire reconsideration of the defence of trade between terminal areas. The whole basis of the old system was the convoy-system, and it now becomes doubtful whether the additional security which convoys afforded is sufficient to outweigh their economical drawbacks and their liability to cause strategic disturbance.

VI.
THE
TWENTIETH
CENTURY

What passing bells for these who die as cattle?
Only the monstrous anger of the guns.
Only the stuttering rifles' rapid rattle
Can patter out their hasty orisons.
No mockeries for them from prayers or bells,
Nor any voice of mourning save the choirs,—
The shrill, demented choirs of wailing shells;
And bugles calling for them from sad shires.
                              —WILFRED OWEN
                    "Anthem for Doomed Youth"

HALFORD MAC KINDER:

FROM

The Geographical Pivot of History

Halford MacKinder was born in 1861 and took up the study of geography at Oxford. In 1899 he became director of the new School of Geography at Oxford and then held a succession of academic appointments, including director of the London School of Economics. In 1899 he made the first ascent of Mount Kenya. In 1910 he was elected to Parliament. From 1919 to 1922 he was British High Commissioner for occupied South Russia. After his return he was knighted and held a number of official appointments. He died in 1947.

An excerpt from the lecture to the Royal Geographical Society in London in 1904. "He was the first," said American Ambassador John Winant during World War II, "to provide us with a global concept of the world." He profoundly influenced "geopolitics."

When historians in the remote future come to look back on the group of centuries through which we are now passing, and see them foreshortened, as we today see the Egyptian dynasties, it may well be that they will describe the last 400 years as the Columbian epoch, and will say that it ended soon after the year 1900. . . .

The all-important result of the discovery of the Cape road to the Indies was to connect the western and eastern coastal navigations of Euro-Asia, even though by a circuitous route, and thus in some measure to neutralize the strategical advantage of the central position of the steppe-nomads by pressing upon them in the rear. The revolution commenced by the great mariners of the Columbian generation endowed Christendom with the widest possible mobility of power, short of a winged mobility. The one and continuous ocean enveloping the divided and insular lands is, of course, the geographical condition of ultimate unity in the command of the sea, and of the whole theory of modern naval strategy as expounded by such writers as Captain Mahan and Mr. Spenser Wilkinson.

The broad political effect was to reverse the relations of Europe and Asia, for whereas in the Middle Ages Europe was caged between an impossible desert to south, an unknown ocean to west, and icy or forested wastes to north and northeast, and in the east and southeast was constantly threatened by the superior mobility of the horsemen and camelmen, Europe now emerged upon the world, multiplying more than thirty-fold the sea surface and coastal lands to which she had access, and wrapping her influence round the Euro-Asiatic land-power which had hitherto threatened her very existence.

As we consider this rapid review of the broader currents of history, does not a certain persistence of geographical relationship become evident? Is not the pivot region of the world's politics that vast area of Euro-Asia which is inaccessible to ships, but in antiquity lay open to the horse-riding nomad, and is today about to be covered with a network of railways? There have been and are here the conditions of a mobility of military and economic power of a far reaching and yet limited character. Russia replaces the Mongol Empire. Her pressure on Finland, on Scandinavia, on Poland, on Turkey, on Persia, on India, and on China replaces the centrifugal raids of the steppe-men.

In the world at large she occupies the central strategical position held by Germany in Europe. She can strike on all sides and be struck from all sides, save the north. The full development of her modern railway mobility is merely a matter of time. Nor is it likely that any possible social revolution will alter her essential relations to the great geographical limits of her existence. Wisely recognising the fundamental limits of her power, her rulers have parted with Alaska; for it is as much a law of policy for Russia to own nothing overseas as for Britain to be supreme on the ocean.

The oversetting of the balance of power in favour of the pivot state, resulting in its expansion over the marginal lands of Euro-Asia, would permit of the use of vast continental resources for fleet-building, and the empire of the world would then be in sight. This might happen if Germany were to ally herself with Russia. The threat of such an event should, therefore, throw France into alliance with the oversea powers, and France, Italy, Egypt, India and Korea would become so many bridge heads where the outside navies would support armies to compel the pivot allies to deploy land forces and prevent them from concentrating their whole strength on fleets. On a smaller scale that was what Wellington accomplished from his seabase at Torres Vedras in the Peninsular War. May not this in the end prove to be the strategical function of India in the British Imperial system?

The westward march of empire appears to me to have been a short rotation of marginal power round the southwestern and western edge of the pivotal area. The Nearer, Middle and Far Eastern questions relate to the unstable equilibrium of inner and outer powers in those parts of the marginal crescent where local power is, at present, more or less negligible.

In conclusion it may be well expressly to point out that the substitution of some new control of the inland area for that of Russia would not tend to reduce the geographical significance of the pivot position. Were the Chinese, for instance, organized by the Japanese, to overthrow the Russian Empire, and conquer its territory, they might constitute the yellow peril to the world's freedom just because they would add an oceanic frontage to the resources of the great continent, an advantage as yet denied to the Russian tenant of the pivot region.

JEAN COLIN:

FROM

Transformation of War

Jean Colin was born in 1864 and after joining the French army became a lecturer at the Ecole de Guerre. His lectures and historical studies, including critical assessments of Napoleon and of the offensive school, provoked considerable controversy. He was promoted general of a brigade after the outbreak of World War I and was killed in 1917.

Colin became a leading critic of the orthodox "Napoleonic" school in France.

It is in battle, the essential act of war, that moral forces act most powerfully and have their preponderant effect. We cannot repeat this too insistently. But whatever we may write about moral forces will now endow with them the man who has none. It is possible to write reams on the part played by decision, ardour, coolness, and all the qualities proper to a leader, but it is not of great profit to do so. . . .

The advantage of the offensive in battle is obvious: it disorganizes the enemy, upsets his plans and combinations; the assailant, to some extent, imposes on him his initiative, his will. And yet of Napoleon's adversaries those who adopted the defensive suffered less grave reverses than those bold persons who opposed their offensive to his. The Moskowa and Waterloo are examples of this. As a matter of fact the law is not the same for all: it is above all necessary that a general should adopt a role proportionate to his capacity, a plan that he feels himself able to follow out methodically amidst dangers, surprise, friction, accidents of all sorts. . . .

The defensive-offensive form succeeded, however, with Wellington in Spain against generals like Soult and Massèna. This enables us to conclude that no exclusive solution can be adopted, and that although we consider the offensive form combined with a wing attack as preferable, we cannot pronounce formally either against frontal attacks or against the defensive. The one essential is to appreciate correctly one's own value and that of one's adversary.

* * *

It is not only the intervention of governments that is to be feared; it is above all the intervention of peoples. This is due to thoughtless passions, and in consequence is usually unreasonable. It imposes unseasonable battles and shameful capitulations.

The numerous and passionate proletariats of great capitals send armies to their ruin, and, above all, it is in their name that armies are sent to their ruin; in their name that a Napoleon III is obliged to remain on the frontier with 240,000 men against 500,000; that a MacMahon is forced to hurl himself into the abyss.

Though the populace does not always impose such disastrous operations, it always assigns an exaggerated importance to the capital. Sometimes, as in 1870–1, it becomes the object of active operations, distracting the attention of generals from what ought to be their only care—victory in the field; sometimes it obliged them to give battle before a capital, instead of postponing the decision.

Far from provoking or exploiting the populace, the duty of political authorities is to pacify, and, if necessary, to suppress popular movements. Once war has begun, the general entrusted with command and possessing the confidence of the nation should act in all freedom.

—L. POPE-HENNESSY
(translator)

ERICH LUDENDORFF:

FROM

The Nation at War

Erich Ludendorff was born in 1865 and, after joining the German army as an infantry officer, became a member of the general staff in 1894. He was largely concerned with the plans for the invasion of France and Belgium. From 1912 to 1914 he commanded an infantry brigade and after the outbreak of war he led the capture of the Belgian fortress of Liege. He then went as chief of staff to Hindenburg on the Eastern Front—Ludendorff being virtually in charge and credited with the victory of Tannenburg over the Russians.

In 1916 when Hindenburg was appointed to nominal supreme military control, Ludendorff effected some improvements, by defensive strategy, on the Western Front. He began to play an increasing part in political affairs, securing the appointment of a new chancellor at home and negotiating the treaty of Brest-Litovsk with the Russians after the Revolution. He had sponsored Lenin's return from Switzerland—and also supported unrestricted submarine warfare.

After the failure of his final offensive in the West and the collapse of Turkey, Ludendorff demanded an immediate surrender by Germany, then attacked the new government and was dismissed.

In 1919 he returned to Germany from Sweden and became the leader of a new nationalist and racialist movement. He supported Hitler in the attempted coup of 1923, was elected to the Reichstag as a national socialist, and later tried to be elected president. He died in 1937.

The nature of totalitarian warfare literally demands the entire strength of the nation, since such a war is directed against it.

And just as the nature of war has changed under the effect of immutable facts and which are legitimate, so the sphere of the tasks incumbent upon politics ought to have widened and the policy itself ought to have changed. Like the totalitarian war, politics, too, must assume a totalitarian character. With a view to the highest output of a nation in a totalitarian war, politics should be the energetic doctrine of the preservation of the people and should carefully consider the requirements and claims of the nation for the preservation of its existence in all spheres of life, and not least in the physical sphere.

War being the highest test of a nation for the preservation of its existence, a totalitarian policy must, for that very reason, elaborate in peace-time plans for the necessary preparations required for the vital struggle of the nation in war, and fortify the foundations for such a vital struggle so strongly that they could not be moved in the heat of war, neither be broken or entirely destroyed through any measures taken by the enemy.

The nature of war has changed, the character of politics has changed, and now the relations existing between politics and the conduct of war must also change. All the theories of Clausewitz should be thrown overboard. Both warfare and politics are meant to serve the preservation of the people, but warfare is the highest expression of the national "will to live" and politics must, therefore, be subservient to the conduct of war.

The declarations of war of the Imperial Chancellor Von Bethmann-Hollweg on Russia and France in August 1914 are still in everybody's memory. They gave the enemy propaganda a useful start and weakened the morale of our people. . . . The German people were further influenced by the circumstances that the German army actually did attack in the West, and so the nation firmly believed that we were waging a war of aggression, which it considered to be identical with a war of conquest. This very soon deprived the nation of the conviction that it must fight for its existence. It was incapable of understanding, nor had it received military training to that effect, that a defensive war such as had been forced upon us, had to be waged by way of attack if we were not to be beaten.

—A. S. RAPPAPORT
(translator)

HERBERT GEORGE WELLS:

FROM

The War in the Air

Herbert George Wells was born in London in 1866 and after an elementary education was apprenticed to a draper. He won a biology scholarship and then became a teacher and journalist. His first attempt at science fiction was a success and for several years he concentrated on this field before turning to realistic novels. He became optimistically involved in the Socialist movement but his increasing disillusionment with human progress was reflected in his last work, published shortly before his death in 1945.

The accidental balance on the side of Progress was far slighter and infinitely more complex and delicate in its adjustments than the people of that time suspected; but that did not alter the fact that it was an effective balance. They did not realize that this age of relative good fortune was an age of immense but temporary opportunity for their kind. They complacently assumed a necessary progress towards which they had no moral responsibility. They did not realize that this security of progress was a thing still to be won or lost, and that the time to win it was a time that passed.

They went about their affairs energetically enough, and yet with a curious idleness towards those threatening things. No one troubled over the real dangers of mankind. They saw their armies and navies grow larger and more portentous; some of their ironclads at the last cost as much as their whole annual expenditure upon advanced education; they accumulated explosives and the machinery of destruction; they allowed their national traditions and jealousies to accumulate; they contemplated a steady enhancement of race hostility as the races drew closer without concern or understanding, and they permitted the growth in their midst of an evil-spirited press, mercenary and unscrupulous, incapable of good and powerful for evil. Their State had practically no control over the press at all. Quite heedlessly they allowed this touch-paper to lie at the door of their war magazine for any spark to fire. The precedents of history were all one tale of the collapse of civilizations, the dangers of the time were manifest. One is incredulous now to believe they could not see.

Could mankind have prevented this disaster of the War in the Air? An idle question that, as idle as to ask could mankind have prevented the decay or the slow decline and fall that turned Assyria and Babylon to empty deserts, the gradual social disorganization, phase by phase, that closed the chapter of the Empire of the West! They could not, because they did not, they had not the will to arrest it.

. . . And at the same time the character of the war altered through the replacement of the huge gas-filled airships by flying-machines as the instruments of war. So soon as the big fleet engagements were over, the Asiatics endeavoured to establish in close proximity to the more vulnerable points of the countries against which they were acting, fortified centres from which flying-machine raids could be made. For a time they had everything their own way in this, and then, as this story has told, the lost secret of the Butteridge machine came to light, and the conflict became equalized and less conclusive than ever. For these small flying-machines, ineffectual for any large expedition or conclusive attack, were horribly convenient for any guerrilla warfare, rapidly and cheaply made, easily used, easily hidden. The design of them was hastily copied and printed in Pinkerville, and scattered broadcast over the United States, and copies were sent to Europe, and there reproduced. Every man, every town, every parish that could, was exhorted to make, and use them. In a little while they were being constructed not only by governments and local authorities, but by robber bands, by insurgent committees, by every type of private person. The peculiar social destructiveness of the Butteridge machine lay in its complete simplicity. It was nearly as simple as a motor-bicycle. The broad outlines of the earlier stages of the war disappeared under its influence, the spacious antagonism of nations and empires and races vanished in a seething mass of detailed conflict. The world passed at a stride from a unity and simplicity broader than that of the Roman Empire at its best, to a social fragmentation as complete as the robber-baron period of the Middle Ages. But this time, for a long descent down gradual slopes of disintegration, comes a fall like a fall over a cliff. Everywhere were men and women perceiving this, and struggling desperately to keep, as it were, a hold upon the edge of the cliff.

A fourth phase follows. Through the struggle against Chaos, in the wake of the Famine, came now another old enemy of humanity—the Pestilence, the Purple Death. But the war does not pause. The flags still fly. Fresh air-fleets rise, new forms of airship, and beneath their swooping struggles the world darkens—scarcely heeded by history.

GIULIO DOUHET:

FROM

The Command of the Air

Giulio Douhet was born in 1869 and joined the Italian army as an artillery officer. From 1912 to 1915 he served as commander of Italy's first aviation unit. He was then transferred to an infantry division. He was court-martialed, retired, and imprisoned for his outspoken criticism of the conduct of the war.

After the Italian defeat at Caporetto in 1917 he was recalled as head of the aviation service and later promoted major-general. After his retirement he devoted himself to expounding his concept of strategic air warfare. He died in 1930.

The Command of the Air, *first published in 1921, was the basis of "the Douhet Doctrine" in Air Forces.*

To have command of the air means to be in a position to prevent the enemy from flying while retaining the ability to fly oneself. Planes capable of carrying moderately heavy loads of bombs already exist, and the construction of enough of them for national defence would not require exceptional resources. The active ingredients of bombs or projectiles, the explosives, the incendiaries, the poison gases, are already being produced. An aerial fleet capable of dumping hundreds of tons of such bombs can easily be organized; therefore, the striking force and magnitude of aerial offensives, considered from the standpoint of either material or moral significance, are far more effective than those of any other offensive yet known.

A nation which has command of the air is in a position to protect its own territory from enemy aerial attack and even to put a halt to the enemy's auxiliary actions in support of his land and sea operations, leaving him powerless to do much of anything. Such offensive actions can not only cut off an opponent's army and navy from their bases of operations, but can also bomb the interior of the enemy's country so devastatingly that the physical and moral resistance of the people would also collapse.

All this is a present possibility, not one in the distant future. And the fact that this possibility exists, proclaims aloud for anyone to understand that to have command of the air is to have *victory*. Without this command, one's portion is defeat and the acceptance of whatever terms the victor is pleased to impose.

Twelve years ago, when the very first aeroplanes began to hedge-hop between field and air, hardly what we would call flying at all today, I began to preach the value of command of the air. From that day to this I have done my level best to call attention to this new weapon of warfare. I argued that the aeroplane should be the third brother of the army and navy. I argued that the day would come when thousands of military planes would ply the air under an independent Ministry of the Air. I argued that the dirigible and other lighter-than-air ships would give way before the superiority of the plane. And everything I argued for then has come true just as I predicted it in 1909.

When, by the exercise of cold logic and mathematical calculation, someone was able to find out the existence of an unknown planet and furnish an astronomer with all the data necessary for its discovery; when by mathematical reasoning the electro-magnetic waves were discovered, thus furnishing Hertz the means with which to carry on his experiments—then we too should have faith in the validity of human reasoning, at least to the extent that the astronomer and Hertz had faith in it. And how much more abstruse their reasonings were than the reasoning I am attempting here!

At this point I ask my readers to stop with me and consider what I have been saying—the arguments are worth while—so that each may come to his own conclusion about it. The problem does not admit of partial solution. It is right or it is not right.

What I have to say is this: In the preparations for national defence we have to follow an entirely new course because the character of future wars is going to be entirely different from the character of past wars.

I say: The World War was only a point on the graph curve showing the evolution of the character of war: at that point the graph curve makes a sharp swerve showing the influence of entirely new factors. For this reason clinging to the past will teach us nothing useful for the future, for that future will be radically different from anything that has gone before. The future must be approached from a new angle.

I say: If these facts are not given careful consideration, the country will have to make great sacrifices in an effort to bring its defence up to date; but even these sacrifices will be of little use, for the defences could not possibly meet the demands of modern military requirements. This can be denied only by refuting my argument.

I ask again: Is it true or is it not true that the strongest army and navy we could muster would be powerless to prevent a determined, well-prepared enemy from cutting them off from their bases of operation and from spreading terror and havoc over the whole country?

We can answer, "No, it is not true," to this question only if we have no intention of providing ourselves with suitable means, in addition to those of the army and navy, with which to meet any such eventuality. But I, for one, have long been answering this question with a categorical "Yes, it is true"; and it is because I am convinced of the imminence of such an eventuality that I have deeply pondered the problem posed by the new forms and weapons of war.

—DINO FERRARI
(translator)

V. I. LENIN:

FROM

War and Socialism;

FROM

Left-Wing Childishness and Petty Bourgeois Morality

Vladimir Ulyanov, who took the name of Lenin, was born in 1870, the son of a Russian schoolmaster. The execution of his brother for terrorism influenced him to devote himself to political agitation and writing. In 1895 he was imprisoned and exiled. After forming the Bolshevik group he returned to Russia in 1905 to take part in the Moscow uprising, but left for Switzerland when it was suppressed. In 1917, after the overthrow of the Tsar, the Germans facilitated his return in order to neutralize the Russian war effort. In October he seized power from the Kerensky government and established the dictatorship of the proletariat. After leading the Communist state through the Civil War and Allied intervention, he suffered a severe stroke in 1922 and died two years later.

War and Socialism *(1915) and* Left-Wing Childishness and Petty Bourgeois Morality *(1919) show the influence of Clausewitz, as well as Marx and Engels.*

FROM *War and Socialism*

Socialists have always condemned wars as barbarous and brutal. Our attitude towards war, however, is fundamentally different from that of bourgeois pacifists (supporters and advocates of peace) and of the anarchists. We differ from the former in that we understand the inevitable connection between wars and class struggles within a country; we understand that wars cannot be abolished unless classes are abolished and socialism is created; we also differ in that we regard civil wars i.e. wars waged by an oppressed class against the oppressing class, by a slave against slave-holders, by serf against landowners, and by wage-earners against the bourgeoisie as fully legitimate, progressive and necessary.

We Marxists differ from both pacifists and anarchists in that we deem it necessary to study each war historically (from the standpoint of Marx's dialectical materialism) and separately. There have been in the past numerous wars which, despite all the horrors, atrocities, distress and suffering that inevitably accompany all wars, were progressive i.e. benefited the development of making by helping to destroy most harmful and reactionary institutions (e.g. an autocracy or serfdom) and the most barbarous despotisms in Europe (the Turkish and Russian).

How, then, can we disclose and define the "substance of a war"? War is the continuation of a policy. Consequently, we must examine the policy pursued prior to the war, the policy that led to and brought about the war. If it was an imperialist policy . . . then the war stemming from that policy is imperialist. If it was a national liberation policy . . . then that war stemming from that policy is a war of national liberation.

FROM *Left-Wing Childishness and Petty Bourgeois Morality*

When we were the representatives of an oppressed class, we did not adopt a frivolous attitude towards defence of the fatherland in an imperialist war. We opposed such defence on principle. Now that we have become the representatives of the ruling class, which has begun to organise socialism, we demand that everybody adopt a serious attitude towards defence of the country. And adopting a serious attitude towards defence of the fatherland in an imperialist war means thoroughly preparing for it and strictly calculating the balance of forces.

If one's forces are obviously small, the best means of defence is retreat into the interior of the country. (Anyone who regards this as an artificial formula, made up to suit the needs of the moment, should read old Clausewitz, one of the greatest authorities on military matters, concerning the lessons of history to be learned in this connection.)

STEPHEN CRANE:

FROM

The Red Badge of Courage

*Stephen Crane was born in New Jersey in 1871 and, after
a brief period at the university, supported himself as a
journalist. With the publication of* The Red Badge of Courage *in 1895 he gained international fame. He went to
Greece and later to Cuba (like Winston Churchill) as a war
correspondent, settled for a time in England, and died in
Germany in 1900.*

A young man growing up in the Civil War . . .

On an incline over which a road wound he saw wild and desperate rushes of men perpetually backward and forward in riotous
surges. These parts of the opposing armies were two long waves that
pitched upon each other madly at dictated points. To and fro they
swelled. Sometimes, one side by its yells and cheers would proclaim
decisive blows, but a moment later the other side would be all yells
and cheers. Once the youth saw a spray of light forms go in
houndlike leaps toward the waving blue lines. There was much howling, and presently it went away with a vast mouthful of prisoners.
Again, he saw a blue wave dash with such thunderous force against a
gray obstruction that it seemed to clear the earth of it and leave nothing but trampled sod. And always in their swift and deadly rushes to
and fro the men screamed and yelled like maniacs.

Particular pieces of fence or secure positions behind collections
of trees were wrangled over, as gold thrones or pearl bedsteads.
There were desperate lunges at these chosen spots seemingly every
instant, and most of them were bandied like light toys between the

contending forces. The youth could not tell from the battle flags flying like crimson foam in many directions which color of cloth was winning.

His emaciated regiment bustled forth with undiminished fierceness when its time came. When assaulted again by bullets, the men burst out in a barbaric cry of rage and pain. They bent their heads in aims of intent hatred behind the projected hammers of their guns. Their ramrods clanged loud with fury as their eager arms pounded the cartridges into the rifle barrels. The front of the regiment was a smoke-wall penetrated by the flashing points of yellow and red.

Wallowing in the fight, they were in an astonishingly short time resmudged. They surpassed in stain and dirt all their previous appearances. Moving to and fro with strained exertion, jabbering the while, they were, with their swaying bodies, black faces, and glowing eyes, like strange and ugly friends jigging heavily in the smoke.

The lieutenant, returning from a tour after a bandage, produced from a hidden receptacle of his mind new and portentous oaths suited to the emergency. Strings of expletives he swung lashlike over the backs of his men, and it was evident that his previous efforts had in nowise impaired his resources.

The youth, still the bearer of the colors, did not feel his idleness. He was deeply absorbed as a spectator. The crash and swing of the great drama made him lean forward, intent-eyed, his face working in small contortions. Sometimes he prattled, words coming unconsciously from him in grotesque exclamations. He did not know that he breathed; that the flag hung silently over him, so absorbed was he.

A formidable line of the enemy came within dangerous range. They could be seen plainly—tall, gaunt men with excited faces running with long strides toward a wandering fence.

At sight of this danger the men suddenly ceased their cursing monotone. There was an instant of strained silence before they threw up their rifles and fired a plumping volley at the foes. There had been no order given; the men, upon recognizing the menace, had immediately let drive their flock of bullets without waiting for word of command.

But the enemy were quick to gain the protection of the wandering line of fence. They slid down behind it with remarkable celerity, and from this position they began briskly to slice up the blue men.

These latter braced their energies for a great struggle. Often, white clinched teeth shone from the dusky faces. Many heads surged to and fro, floating upon a pale sea of smoke. Those behind the fence frequently shouted and yelped in taunts and gibelike cries, but the regiment maintained a stressed silence. Perhaps, at this new assault the men recalled the fact that they had been named mud diggers, and it made their situation thrice bitter. They were breathlessly intent upon keeping the ground and thrusting away the rejoicing body of the enemy. They fought swiftly and with a despairing savageness denoted in their expressions.

The youth had resolved not to budge whatever should happen. Some arrows of scorn that had buried themselves in his heart had generated strange and unspeakable hatred. It was clear to him that his final and absolute revenge was to be achieved by his dead body lying, torn and gluttering, upon the field. This was to be a poignant retaliation upon the officer who had said "mule drivers," and later "mud diggers," for in all the wild graspings of his mind for a unit responsible for his sufferings and commotions he always seized upon the man who had dubbed him wrongly. And it was his idea, vaguely formulated, that his corpse would be for those eyes a great and salt reproach.

MARCEL PROUST:

FROM

Remembrance of Things Past

Marcel Proust was born in 1871, of a prosperous French Jewish family. He did his military service at Orleans—and applied unsuccessfully for an extension—then studied law and literature. After participating in Parisian social life he began to withdraw at the time of the Dreyfus affair, about which he felt strongly. From 1905 he devoted his life in se-clusion to his great work, Remembrance of Things Past, *which was published in successive volumes until his death, after long illness, in 1922.*

In The Guermantes Way *the narrator visits his friend, Saint-Loup at a garrison where the latter is a trooper.*

What gave me pleasure to-day would not, perhaps, leave me indifferent to-morrow, as had always happened hitherto; the creature that I still was at this moment was not, perhaps, doomed to imme-diate destruction, since to the ardent and fugitive passion which I had felt on these few evenings for everything connected with military life, Saint-Loup, by what he had just been saying to me, touching the art of war, added an intellectual foundation, of a permanent character, capable of attaching me to itself so strongly that I might, without any attempt to deceive myself, feel assured that after I had left Doncières I should continue to take an interest in the work of my friends there, and should not be long in coming to pay them another visit. At the same time, so as to make quite sure that this art of war was indeed an art in the true sense of the word:

"You interest me—I beg your pardon, *tu* interest me enormously," I said to Saint-Loup, "but tell me, there is one point that puzzles me. I feel that I could be keenly thrilled by the art of strategy, but if so I must first be sure that it is not so very different from the other arts, that knowing the rules is not everything. You tell me that plans of battles are copied. I do find something aesthetic, just as you said, in seeing beneath a modern battle the plan of an older one, I can't tell you how attractive it sounds. But then, does the genius of the commander count for nothing? Does he really do no more than apply the rules? Or, in point of science, are there great generals as there are great surgeons, who, when the symptoms exhibited by two states of ill-health are identical to the outward eye, nevertheless feel, for some infinitesimal reason, founded perhaps on their experience, but interpreted afresh, that in one case they ought to do one thing, in another case another; that in one case it is better to operate, in another to wait?"

"I should just say so! You will find Napoleon not attacking when all the rules ordered him to attack, but some obscure divination warned him not to. For instance, look at Austerlitz, or in 1806 take his instructions to Lannes. But you will find certain generals slavishly imitating one of Napoleon's movements and arriving at a diametrically opposite result. There are a dozen examples of that in 1870. But even for the interpretation of what the enemy *may* do, what he actually does is only a symptom which may mean any number of different things. Each of them has an equal chance of being the right thing, if one looks only to reasoning and science, just as in certain difficult cases all the medical science in the world will be powerless to decide whether the invisible tumour is malignant or not, whether or not the operation ought to be performed. It is his instinct, his divination—like Mme. de Thèbes (you follow me?)—which decides, in the great general as in the great doctor. Thus I've been telling you, to take one instance, what might be meant by a reconnaissance on the eve of a battle. But it may mean a dozen other things also, such as to make the enemy think you are going to attack him at one point whereas you intend to attack him at another, to put out a screen which will prevent him from seeing the preparations for your real operation, to force him to bring up fresh troops, to hold them, to immobilise them in a different place from where they are needed, to form an estimate of the forces at his disposal, to feel him, to force him to shew his hand. Sometimes, indeed, the fact that you employ an immense number of troops in an operation is by no means a proof that that is your true objective; for you may be justified in carrying it out, even if it is only a feint, so that your feint may have a better chance of deceiving the enemy. If I had time now to go through the Napoleonic wars from this point of view, I assure you that these simple classic movements which we study here, and which you will come and see us practising in the field, just for the pleasure of a walk, you young rascal—no, I know you're not well, I apologise!—well, in a war, when you feel behind you the vigilance, the judgment, the profound study of the Higher Command, you are as much moved by them as by the simple lamps of a lighthouse, only a material combustion, but an emanation of the spirit, sweeping through space to warn ships of danger. I may have been wrong, perhaps, in speaking to you only of the literature of war. In reality, as the formation of the soil, the direction of wind and light tell us which way a tree will grow, so the conditions in which a campaign is fought, the features of the country through which you march, prescribe, to a certain extent, and limit the number of the plans among which the general has to choose. Which means that along a mountain range, through a system of valleys, over certain plains, it is almost with the inevitability and the tremendous beauty of an avalanche that you can forecast the line of an army on the march."

"Now you deny me that freedom of choice in the commander, that power of divination in the enemy who is trying to discover his plan, which you allowed me a moment ago."

"Not at all. You remember that book of philosophy we read together at Balbec, the richness of the world of possibilities compared with the real world. Very well. It is the same again with the art of strategy. In a given situation there will be four plans that offer themselves, one of which the general has to choose, as a disease may pass through various phases for which the doctor has to watch. And here again the weakness and greatness of the human elements are fresh causes of uncertainty. For of these four plans let us assume that contingent reasons (such as the attainment of minor objects, or time, which may be pressing, or the smallness of his effective strength and shortage of rations) lead the general to prefer the first, which is less perfect, but less costly also to carry out, is more rapid, and has for its terrain a richer country for feeding his troops. He may, after having begun with this plan, which the enemy, uncertain at first, will soon detect, find that success lies beyond his grasp, the difficulties being too great (that is what I call the element of human weakness), abandon it and try the second or third or fourth. But it may equally be that he has tried the first plan (and this is what I call human greatness) merely as a feint to pin down the enemy, so as to surprise him later at a point where he has not been expecting an attack. Thus at Ulm, Mack, who expected the enemy to advance from the west, was surrounded from the north where he thought he was perfectly safe. My example is not a very good one, as a matter of fact. And Ulm is a better type of enveloping battle, which the future will see reproduced, because it is not only a classic example from which generals will seek inspiration, but a form that is to some extent necessary (one of several necessities, which leaves room for choice, for variety) like a type of crystallisation. But it doesn't much matter, really, because these conditions are after all artificial. To go back to our philosophy book; it is like the rules of logic or scientific laws, reality does conform to it more or less, but bear in mind that the great mathematician Poincaré is by no means certain that mathematics are strictly accurate. As to the rules themselves, which I mentioned to you, they are of secondary importance really, and besides they are altered from time to time. We cavalrymen, for instance, have to go by the *Field Service* of 1895, which, you may say, is out of date since it is based on the old and obsolete doctrine which maintains that cavalry warfare has little more than a moral effect, in the panic that the charge creates in the enemy. Whereas the more intelligent of our teachers, all the best brains in the cavalry, and particularly the major I was telling you about, anticipate on the contrary that the decisive victory will be obtained by a real hand to hand encounter in which our weapons will be sabre and lance and the side that can hold out longer will win, not simply morally and by creating panic, but materially."

"Saint-Loup is quite right, and it is probable that the next *Field Service* will shew signs of this evolution," put in my other neighbour.

"I am not ungrateful for your support, for your opinions seem to make more impression upon my friend than mine," said Saint-Loup with a smile, whether because the growing attraction between his comrade and myself annoyed him slightly or because he thought it graceful to solemnise it with this official confirmation. "Perhaps I may have underestimated the importance of the rules; I don't know. They do change, that must be admitted. But in the mean time they control the military situation, the plans of campaign and concentration. If they reflect a false conception of strategy they may be the principal cause of defeat. All this is a little too technical for you," he

remarked to me. "After all, you may say that what does most to ac-
celerate the evolution of the art of war is wars themselves. In the
course of a campaign, if it is at all long, you will see one belligerent
profiting by the lessons furnished him by the successes and mistakes,
perfecting the methods of the other, who will improve on him in turn.
But all that is a thing of the past. With the terrible advance of artillery, the wars of the future, if there are to be any more wars, will be
so short that, before we have had time to think of putting our lessons
into practice, peace will have been signed."

–C. K. SCOTT-MONCRIEFF
(translator)

WINSTON CHURCHILL:

FROM

The River War

Winston Churchill was born at Blenheim Palace in England in 1874, a descendant of the great Duke of Marlborough and half-American. After attending the Royal Military College he joined a cavalry regiment in 1895 and reported the same year on the Cuban War of Independence. Then he served in India, with the Nile Expeditionary Force, engaged at the Battle of Omdurman and in the Boer War in South Africa, where he was captured. After writing on these experiences, he entered politics. A variety of political allegiances and offices culminated in his leadership during World War II as prime minister of Britain from 1940 to 1945. After resigning from the premiership for the second time, he continued to sit as a member of Parliament and to write history. He died in 1965.

The British war against the Sudanese (or Dervishes) ended with the defeat of their leader, the Khalifa, at Omdurman in 1898. There the author took part in what has often been described as the last cavalry charge in history.

The crest of the ridge was only half a mile away. It was found unoccupied. The rocky mass of Surgham obstructed the view and concealed the great reserve collected around the Black Flag. But southward, between us and Omdurman, the whole plain was exposed. It was infested with small parties of Dervishes, moving about, mounted and on foot, in tens and twenties. Three miles away a broad stream of fugitives, of wounded, and of deserters flowed from the Khalifa's army to the city. The sight was sufficient to excite the fierc-

est instincts of cavalry. Only the scattered parties in the plain appeared to prevent a glorious pursuit.

The signalling officer was sent to heliograph back to the Sirdar
that the ridge was unoccupied and that several thousand Dervishes
could be seen flying into Omdurman. Pending the answer, we waited;
and looking back northwards, across the front of the *zeriba,* where
the first attack had been stopped, perceived a greyish-white smudge,
perhaps a mile long. The glass disclosed details—hundreds of tiny
white figures heaped or scattered; dozens hopping, crawling, staggering away; a few horses standing stolidly among the corpses; a few un-
wounded men dragging off their comrades. Then the heliograph in
the *zeriba* began to talk in flashes of light that opened and shut
capriciously. The actual order is important. "Advance," said the
helio, "and clear the left flank, and use every effort to prevent the
enemy re-entering Omdurman." That was all, but it was sufficient.

But all this time the enemy had been busy. At the beginning of
the battle the Khalifa had posted a small force of 700 men on his extreme right, to prevent his line of retreat to Omdurman being
harassed. This detachment was composed entirely of the Hadendoa
tribesmen of Osman Digna's flag, and was commanded by one of his
subordinate Emirs, who selected a suitable position in the shallow
khor. As soon as the 21st Lancers left the *zeriba* the Dervish scouts
on the top of Surgham carried the news to the Khalifa. It was said
that the English cavalry were coming to cut him off from Omdurman.
Abdullah thereupon determined to strengthen his extreme right; and
he immediately ordered four regiments, each 500 strong, drawn from
the force around the Black Flag and under the Emir Ibrahim Khalil,
to reinforce the Hadendoa in the *khor.*

We advanced at a walk in mass for about 300 yards. The scattered parties of Dervishes fell back and melted away, and only one
straggling line of men in dark blue waited motionless a quarter of a
mile to the left front. They were scarcely a hundred strong. The regiment formed into line of squadron columns, and continued at a walk
until within 300 yards of this small body of Dervishes. There was
complete silence, intensified by the recent tumult. Far beyond the
thin blue row of Dervishes the fugitives were visible streaming into
Omdurman. And should these few devoted men impede a regiment?

Yet it were wiser to examine their position from the other flank
before slipping a squadron at them. The heads of the squadrons
wheeled slowly to the left, and the Lancers, breaking into a trot,

began to cross the Dervish front in column of troops. Thereupon and with one accord the blue-clad men dropped on their knees, and there burst out a loud crackling fire of musketry. It was hardly possible to miss such a target at such a range. Horses and men fell at once. The only course was plain and welcome to all. The Colonel, nearer than his regiment, already saw what lay behind the skirmishers. He ordered "Right wheel into line" to be sounded. The trumpet jerked out a shrill note, heard faintly above the trampling of the horses and the noise of the rifles. On the instant all the sixteen troops swung round and locked up into a long galloping line, and the 21st Lancers were committed to their first charge in war.

Two hundred and fifty yards away the dark-blue men were firing madly in a thin film of light-blue smoke. Their bullets struck the hard gravel into the air, and the troopers, to shield their faces from the stinging dust, bowed their helmets forward, like the Cuirassiers at Waterloo. The pace was fast and the distance short. Yet, before it was half covered, the whole aspect of the affair changed. A deep crease in the ground—a dry watercourse, a *khor*—appeared where all had seemed smooth, level plain; and from it there sprang, with the suddenness of a pantomime effect and a high-pitched yell, a dense white mass of men nearly as long as our front and about twelve deep. A score of horsemen and a dozen bright flags rose as if by magic from the earth. Eager warriors sprang forward to anticipate the shock. The rest stood firm to meet it. The Lancers acknowledged the apparition only by an increase of pace. Each man wanted sufficient momentum to drive through such a solid line. The flank troops, seeing that they overlapped, curved inwards like the horns of a moon. But the whole event was a matter of seconds. The riflemen, firing bravely to the last, were swept head over heels into the *khor,* and jumping down with them, at full gallop and in the closest order, the British squadrons struck the fierce brigade with one loud furious shout.

The collision was prodigious. Nearly thirty Lancers, men and horses, and at least two hundred Arabs were overthrown. The shock was stunning to both sides, and for perhaps ten wonderful seconds no man heeded his enemy. Terrified horses wedged in the crowd, bruised and shaken men, sprawling in heaps, struggled, dazed and stupid, to their feet, panted, and looked about them. Several fallen Lancers had even time to remount. Meanwhile the impetus of the cavalry carried them on.

Stubborn and unshaken infantry hardly ever meet stubborn and unshaken cavalry. Either the infantry run away and are cut down in flight, or they keep their heads and destroy nearly all the horsemen by their musketry. On this occasion two living walls had actually crashed together. The Dervishes fought manfully. They tried to hamstring the horses. They fired their rifles, pressing the muzzles into the very bodies of their opponents. They cut reins and stirrup-leathers. They flung their throwing-spears with great dexterity. They tried every device of cool, determined men practised in war and familiar with cavalry; and, besides, they swung sharp, heavy swords which bit deep. The hand-to-hand fighting on the further side of the *khor* lasted for perhaps one minute. Then the horses got into their stride again, the pace increased, and the Lancers drew out from among their antagonists. Within two minutes of the collision every living man was clear of the Dervish mass. All who had fallen were cut at with swords till they stopped quivering.

Two hundred yards away the regiment halted, rallied, faced about, and in less than five minutes were re-formed and ready for a second charge. The men were anxious to cut their way back through their enemies. We were alone together—the cavalry regiment and the Dervish brigade. The ridge hung like a curtain between us and the army. The general battle was forgotten, as it was unseen. This was a private quarrel. The other might have been a massacre; but here the fight was fair, for we too fought with sword and spear. Indeed the advantage of ground and numbers lay with them. All prepared to settle the debate at once and for ever. But some realization of the cost of our wild ride began to come to those who were responsible. Riderless horses galloped across the plain. Men, clinging to their saddles, lurched helplessly about, covered with blood from perhaps a dozen wounds. Horses, streaming from tremendous gashes, limped and staggered with their riders. In 120 seconds five officers, 65 men, and 119 horses out of fewer than 400 had been killed or wounded.

The Dervish line, broken by the charge, began to re-form at once. They closed up, shook themselves together, and prepared with constancy and courage for another shock. But on military considerations it was desirable to turn them out of the *khor* first and thus deprive them of their vantage-ground. The regiment again drawn up, three squadrons in line and the fourth in column, now wheeled to the right, and, galloping round the Dervish flank, dismounted and opened a heavy fire with their magazine carbines. Under the pressure of this fire the enemy changed front to meet the new attack, so that both sides were formed at right angles to their original lines. When the Dervish change of front was completed, they began to advance against the dismounted men. But the fire was accurate, and there can be little doubt that the moral effect of the charge had been very great, and that these brave enemies were no longer unshaken. Be this as it may, the fact remains that they retreated swiftly, though in good order, towards the ridge of Surgham Hill, where the Khalifa's Black Flag still waved, and the 21st Lancers remained in possession of the ground—and of their dead.

JOHN FREDERICK CHARLES FULLER:

FROM

Memoirs of an Unconventional Soldier

John Frederick Charles Fuller was born in 1878 and, after only a year at the Royal Military College, joined an infantry regiment in 1898. He served in the Boer War until 1902 and then in India where he became deeply interested in Yoga philosophy—about which he wrote a book. After his return to England he published several small works on infantry training. At the outbreak of World War I he went to the Western Front as a staff officer and was attached to the Heavy Branch of the Machine Gun Corps—the embryo tank formation. In 1916 he became Chief General Staff Officer of the Tank Corps and planned the surprise tank attack at Cambrai in 1917, the first armored warfare engagement in history. He returned to England to help the Imperial General Staff double the size of the Tank Corps.

In 1922 he became chief instructor at the Staff College and in 1926 military assistant to the Chief of the Imperial General Staff. He was then selected to command an experimental force with armored units, but this proved abortive and, after commanding infantry brigades, he prematurely retired from the army in 1930 as a major-general.

He attended German army maneuvers at Hitler's invitation in 1936 and stood for Parliament in Britain, unsuccessfully, as a member of the British Fascist party. He continued to write numerous books on military history and science. He died in 1966.

Fuller included in his Memoirs *the text of "Plan 1919"—
possibly the first military formulation of the* blitzkrieg.

Though of slow development, this idea suddenly flashed across
my mind during our debacle in March. What did I then see? Tens of
thousands of our men being pulled back by their panic-stricken
Headquarters. I saw Army Headquarters retiring, then Corps, next
Divisional and lastly Brigade. I saw the intimate connection between
will and action, and that action without will loses all co-ordination:
that without an active and directive brain, an army is reduced to a
mob. Then I realised that if this idea could be rationalised—by which
I mean taken out of the realms of chaos and digested scientifically—a
new tactic could be evolved, which would enable a comparatively
small tank army to fight battles like Issus and Arbela over again.
What was the secret of these engagements? It was that whilst Alexander's phalanx held the enemy's battle body in a clinch, he and his
Companion Cavalry struck at the enemy's will, concentrated as it was
in the person of Darius. Once this will was paralysed, the body became inarticulate.

On May 24 I elaborated this idea in "Plan 1919," which, like
so many of my tactical papers, was a kind of military novelette. Here
I will give it almost word for word as it was written, condensing it
only in a few places, and improving on its hasty grammar. Later on I
rewrote it in clearer form; yet here, I think, it is more honest to quote
the first edition; for it was this edition which was considered by Sir
Henry Wilson, General Harington, Mr. Churchill and others at the
time. It reads as follows:

"(1) *The Influence of Tanks on Tactics:* Tactics, or the art of
moving armed men on the battlefield, change according to the weapons used and the means of transportation. Each new or improved
weapon or method of movement demands a corresponding change in
the art of war, and to-day the introduction of the tank entirely revolutionises this art in that:

"(i) It increases mobility by replacing muscular by mechanical
power.

"(ii) It increases security by using armour plate to cut out the
bullet.

"(iii) It increases offensive power by relieving the soldier from
having to carry his weapons, and the horse from having to haul them,

and it multiplies the destructive power of weapons by increasing ammunition supply.

"Consequently, petrol enables an army to obtain greater effect from its weapons, in a given time and with less loss to itself than an army which relies upon muscular energy. Whilst securing a man dynamically, it enables him to fight statically; consequently, it superimposes naval upon land tactics; that is, it enables men to discharge their weapons from a moving platform protected by a fixed shield.

"(2) *The Influence of Tanks on Strategy:* Strategy is woven upon communications; hitherto upon roads, railways, rivers and canals. To-day the introduction of a cross-country petrol-driven machine, tank or tractor, has expanded communications to include at least 75 per cent of the theatre of war over and above communications as we at present know them. The possibility to-day of maintaining supply and of moving weapons and munitions over the open, irrespective of roads and without the limiting factor of animal endurance, introduces an entirely new problem in the history of war. At the moment he who grasps the full meaning of this change, namely, that the earth has now become as easily traversable as the sea, multiplies his chances of victory to an almost unlimited extent. Every principle of war becomes easy to apply if movement can be accelerated and accelerated at the expense of the opposing side. To-day, to pit an overland mechanically moving army against one relying on roads, rails and muscular energy is to pit a fleet of modern battleships against one of wind-driven three-deckers. The result of such an action is not even within the possibilities of doubt; the latter will for a certainty be destroyed, for the highest form of machinery must win, because it saves time and time is the controlling factor in war.

"(3) *The Present Tank Tactical Theory:* Up to the present the theory of the tactical employment of tanks has been based on trying to harmonise their powers with existing methods of fighting, that is, with infantry and artillery tactics. In fact, the tank idea, which carries with it a revolution in the methods of waging war, has been grafted on to a system it is destined to destroy, in place of being given free scope to develop on its own lines. This has been unavoidable, because of the novelty of the idea, the uncertainty of the machine and ignorance in its use.

"Knowledge can best be gained by practical experience, and at first this experience is difficult to obtain unless the new idea is grafted to the old system of war. Nevertheless, it behoves us not to forget that the tank (a weapon as different from those which preceded it as the armoured knight was from the unarmoured infantry who preceded him) will eventually, as perfection is gained and numbers are increased, demand a fundamental change in our tactical theory of battle.

"The facts upon which this theory is based are now rapidly changing, and unless it changes with them, we shall not develop to the full the powers of the new machine; that is, the possibility of moving rapidly in all directions with comparative immunity to small-arm fire.

"From this we can deduce the all-important fact that infantry, as at present equipped, will become first a subsidiary and later on a useless arm on all ground over which tanks can move. This fact alone revolutionises our present conception of war, and introduces a new epoch in tactics.

"(4) *The Strategical Objective:* Irrespective of the arm employed, the principles of strategy remain immutable, changes in weapons affecting their application only. The first of all strategical principles is 'the principle of the object,' the object being 'the destruction of the enemy's fighting strength.' This can be accomplished in several ways, the normal being the destruction of the enemy's field armies—his fighting personnel.

"Now, the potential fighting strength of a body of men lies in its organisation; consequently, if we can destroy this organisation, we shall destroy its fighting strength and so have gained our object.

"There are two ways of destroying an organisation:

"(i) By wearing it down (dissipating it).

"(ii) By rendering it inoperative (unhinging it).

"In war the first comprises the killing, wounding, capturing and disarming of the enemy's soldiers—body warfare. The second, the rendering inoperative of his power of command—brain warfare. Taking a single man as an example: the first method may be compared to a succession of slight wounds which will eventually cause him to bleed to death; the second—a shot through the brain.

"The brains of an army are its Staff—Army, Corps and Divisional Headquarters. Could we suddenly remove these from an extensive sector of the German front, the collapse of the personnel they control would be a mere matter of hours, even if only slight opposition were put up against it. Even if we put up no opposition at all, but in addition to the shot through the brain we fire a second shot through the stomach, that is, we dislocated the enemy's supply system behind his protective front, his men will starve to death or scatter.

"Our present theory, based on our present weapons, weapons of limited range of action, has been one of attaining our strategical object by brute force; that is, the wearing away of the enemy's muscles, bone and blood. To accomplish this rapidly with tanks will demand many thousands of these machines, and there is little likelihood of our obtaining the requisite number by next year; therefore let us search for some other means, always remembering that probably, at no time in the history of war, has a difficulty arisen the solution of which has not at the time in question existed in some man's head, and frequently in those of several. The main difficulty has nearly always lurked, not in the solution itself, but in its acceptance by those who have vested interests in the existing methods.

"As our present theory is to destroy 'personnel,' so should our new theory be to destroy 'command,' not after the enemy's personnel has been disorganised, but before it has been attacked, so that it may be found in a state of complete disorganisation when attacked. Here we have the highest application of the principle of surprise—surprise by novelty of action, or the impossibility of establishing security even when the unexpected has become the commonplace.

"(5) *The Suggested Solution:* In order to render inoperative the Command of the German forces on any given front, what are the requirements?

"From the German front line the average distance to nine of their Army Headquarters is eighteen miles; to three Army Group Headquarters forty-five miles; and the distance away of their Western G.H.Q. is one hundred miles. For purposes of illustration the eighteen-mile belt or zone containing Army, Corps and Divisional Headquarters will prove sufficient.

"Before reaching these Headquarters elaborate systems of trenches and wire entanglements, protected by every known type of missile-throwing weapon, have to be crossed.

"To penetrate or avoid this belt of resistance, which may be compared to a shield protecting the system of command, two types of weapons suggest themselves:

"(i) The aeroplane.

"(ii) The tank.

"The first is able to surmount all obstacles; the second to traverse most.

"The difficulties in using the first are very great; for even if landing-grounds can be found close to the various Headquarters, once the men are landed, they are no better armed than the men they will meet; in fact, they may be compared to dismounted cavalry facing infantry.

"The difficulties of the second are merely relative. At present we do not possess a tank capable of carrying out the work satisfactorily, yet this is no reason why we should not have one nine months hence if all energies are devoted to design and production. The idea of such a tank exists, and it has already been considered by many good brains; it is known as the 'medium D tank,' and its specifications are as follows:

"(i) To move at a maximum speed of 20 miles an hour.

"(ii) To possess a circuit of action of 150 to 200 miles.

"(iii) To be able to cross a 13- to 14-foot gap.

"(iv) To be sufficiently light to cross ordinary road, river and canal bridges.

"(6) *The Tactics of the Medium D Tank:* The tactics of the Medium D tank are based on the principles of movement and surprise, its tactical object being to accentuate surprise by movement, not so much through rapidity as by creating unexpected situations. We must never do what the enemy expects us to do; instead, we must mislead him, that is, control his brain by our own. We must suggest to him the probability of certain actions, and then, when action is demanded, we must develop it in a way diametrically opposite to the one we have suggested through our preparations.

"Thus, in the past, when we massed men and guns opposite a given sector, he did the same and frustrated our attack by making his own defences so strong that we could not break through them, or if we did, were then too exhausted to exploit our initial success. At the battle of Cambrai, when our normal method was set aside, our blow could not be taken advantage of, because the forces which broke through were not powerful enough to cause more than local disorganisation. The enemy's strength was not in his front line, but in rear of it; we could not, in the circumstances which we had not created, disorganise his reserves. Reserves are the capital of victory.

"A study of Napoleon's tactics will show us that the first step he took in battle was not to break his enemy's front, and then when his forces were disorganised risk being hit by the enemy's reserves; but instead to draw the enemy's reserves into the fire fight, and directly they were drawn in to break through them or envelop them. Once this was done, security was gained; consequently, a pursuit could be carried out, a pursuit being more often than not initiated by troops disorganised by victory against troops disorganised by defeat."

LEON TROTSKY:

FROM

History of the Russian Revolution

*Leon Trotsky was born in the Ukraine in 1879, the son of a
Jewish farmer. He went to the university to study mathematics but soon left to become a professional revolutionary.
Banished to Siberia in 1900, he escaped abroad and joined
Lenin—but soon broke with him and joined the Menshevik
faction. During the 1905 uprising he returned to Russia and
led the St. Petersburg Soviet. Banished a second time to
Siberia, he again escaped abroad and continued in fac-
tional disputes, advocating permanent worldwide revolution in opposition to Lenin and other "conservatives."*

*After the overthrow of the Tsar he returned to Russia
from the United States and became chairman of the Petrograd Soviet, setting up the Military Revolutionary Committee which organized the Bolshevik seizure of power in the
capital in October 1917. Commissar for foreign affairs in
the new Communist government, he led the peace negotiations with the Germans which took Russia out of the
war and imposed enormous losses of territory on her.
From 1918 he was commissar for military and naval affairs,
creating the Red Army and organizing victory in the Civil
War.*

*After Lenin's death in 1924 he lost out in the struggle
for power to his main antagonist, Stalin, and was banished
to central Asia in 1928. The following year he was expelled
from the Soviet Union and lived as an exile in many countries, denouncing Stalin and writing numerous works on the
history and theory of communism. Most of Trotsky's associates, as well as most of his former opponents, were
liquidated in the great purges of the thirties. Trotsky himself was killed by a Stalinist agent in Mexico in 1940.*

In "The Storming of the Winter Palace" *Trotsky describes
and analyzes the Bolshevik seizure of power in 1917, when
he took charge of military operations.*

It had been proposed in the preliminary calculations to occupy
the Winter Palace on the night of the 25th, at the same time with the
other commanding high points of the capital. It was finally decided to
surround the region of the palace with an uninterrupted oval, the
longer axis of which should be the quay of the Neva. On the riverside
the circle should be closed up by the Peter and Paul fortress, the
Aurora, and other ships summoned from Kronstadt and the navy. In
order to prevent or paralyse the attempts to strike at the rear with
Cossacks and junker detachments, it was decided to establish imposing flank defences composed of revolutionary detachments.

The plan as a whole was too heavy and complicated for the
problem it aimed to solve. The time allotted for preparation proved
inadequate. Small inco-ordinations and omissions came to light at
every step, as might be expected. In one place the direction was in-
correctly indicated, in another the leader came late, having misread
the instructions; in a third they had to wait for a rescuing armoured
car. To call out the military units, unite them with the Red Guards,
occupy the fighting positions, make sure of communications among
them all and with headquarters—all this demanded a good many
hours more than had been imagined by the leaders quarrelling over
their map of Petrograd.

When the Military Revolutionary Committee announced at
about ten o'clock in the morning that the government was
overthrown, the extent of this delay was not yet clear even to those in
direct command of the operation. Podvoisky had promised the fall of
the palace "not later than twelve o'clock." Up to that time everything
had run so smoothly on the military side that nobody had any reason
to question the hour. But at noon it turned out that the besieging
force was still not filled out, the Kronstadters had not arrived, and
that meanwhile the defence of the palace had been reinforced. This
loss of time, as almost always happens, made new delays necessary.

Under urgent pressure from the Committee the seizure of the
palace was now set for three o'clock—and this time "conclusively."
Counting on this new decision, the spokesman of the Military Revolutionary Committee expressed to the afternoon session of the Soviet

the hope that the fall of the Winter Palace would be a matter of the next few minutes. But another hour passed and brought no decision. Podvoisky, himself in a state of white heat, asserted over the telephone that by six o'clock the palace would be taken no matter what it cost. His former confidence, however, was lacking. And indeed the hour of six did strike and the denouement had not begun. Beside themselves with the urgings of Smolny, Podvoisky and Antonov now refused to set any hour at all. That caused serious anxiety.

Politically it was considered necessary that at the moment of the opening of the Congress the whole capital should be in the hands of the Military Revolutionary Committee: That was to simplify the task of dealing with the opposition at the Congress, placing them before an accomplished fact. Meanwhile the hour appointed for opening the Congress had arrived, had been postponed, and arrived again, and the Winter Palace was still holding out. Thus the siege of the palace, thanks to its delay, became for no less than twelve hours the central problem of the insurrection.

Antonov-Ovseenko had agreed with Blagonravov that after the encirclement of the palace was completed, a red lantern should be raised on the flagpole of the fortress. At this signal the *Aurora* would fire a blank volley in order to frighten the palace. In case the besieged were stubborn the fortress should begin to bombard the palace with real shells from the light guns. If the palace did not surrender even then, the *Aurora* would open a real fire from its six-inch guns. The object of this gradation was to reduce to a minimum the victims and the damage, supposing they could not be altogether avoided. But the too complicated solution of a simple problem threatened to lead to an opposite result. The difficulty of carrying this plan out is too obvious. They are to begin with a red lantern: It turns out that they have none on hand. They lose time hunting for it, and finally find it. However, it is not so simple to tie a lantern to a flagpole in such a way that it will be visible in all directions. Efforts are renewed and twice renewed with a dubious result, and meanwhile the precious time is slipping away.

The chief difficulty developed, however, in connection with the artillery. According to a report made by Blagonravov the bombardment of the capital had been possible on a moment's notice ever since noon. In reality it was quite otherwise. Since there was no permanent artillery in the fortress, except for that rusty-muzzled cannon which announces the noon hour, it was necessary to lift field guns up to the fortress walls. That part of the programme had actually been carried out by noon. But a difficulty arose about finding gunners. It had been known in advance that the artillery company—one of those which had not come out on the side of the Bolsheviks in July—was hardly to be relied on. Only the day before it had meekly guarded a bridge under orders from headquarters. A blow in the back was not to be expected from it, but the company had no intention of going through fire for the Soviets. When the time came for action the ensign reported: The guns are rusty; there is no oil in the compressors; it is impossible to shoot. Very likely the guns really were not in shape, but that was not the essence of it. The artillerists were simply dodging the responsibility, and leading the inexperienced commissars by the nose. Antonov dashes up on a cutter in a state of fury. Who is sabotaging the plan? Blagonravov tells him about the lantern, about the oil, about the ensign. They both start to go up to the cannon. Night, darkness, puddles in the court from the recent rains. From the other side of the river comes hot rifle fire and the rattle of machine-guns. In the darkness Blagonravov loses the road. Splashing through the puddles, burning with impatience, stumbling and falling in the mud, Antonov blunders after the commissar through the dark court. "Beside one of the weakly glimmering lanterns," relates Blagonravov . . . "Antonov suddenly stopped and peered inquiringly at me over his spectacles, almost touching my face. I read in his eyes a hidden alarm." Antonov had for a second suspected treachery where there was only carelessness.

The position of the guns was finally found. The artillery men were stubborn: Rust. . . Compressors. . . Oil. Antonov gave orders to bring gunners from the naval polygon and also to fire a signal from the antique cannon which announced the noon hour. But the artillery men were suspiciously long monkeying with the signal cannon. They obviously felt that the commanders too, when not far-off at the telephone but right beside them, had no firm will to resort to heavy artillery. Even under the very clumsiness of this plan for artillery fire the same thought is to be left lurking: Maybe we can get along without it.

Somebody is rushing through the darkness of the court. As he comes near he stumbles and falls in the mud, swears a little but not angrily, and then joyfully and in a choking voice cries out: "The palace has surrendered and our men are there." Rapturous embraces. How lucky there was a delay! "Just what we thought!" The compres-

sors are immediately forgotten. But why haven't they stopped shooting on the other side of the river? Maybe some individual groups of
junkers are stubborn about surrendering. Maybe there is a misunderstanding? The misunderstanding turned out to be good news: not
the Winter Palace was captured, but only the general staff. The siege
of the palace continued.

The palace still holds out. It is time to have an end. The order is
given. Firing begins—not frequent and still less effectual. Out of
thirty-five shots fired in the course of an hour and a half or two
hours, only two hit the mark, and they can only injure the plaster.
The other shells go high, fortunately not doing any damage in the
city. Is lack of skill the real cause? They were shooting across the
Neva with a direct aim at a target as impressive as the Winter Palace:
that does not demand a great deal of artistry. Would it not be truer to
assume that even Lashevitch's artillerymen intentionally aimed high
in the hope that things would be settled without destruction and
death? It is very difficult now to hunt out any trace of the motive
which guided the two nameless sailors. Have they dissolved in the im-
measurable Russian land, or, like so many of the October fighters,
did they lay down their heads in the civil wars of the coming months
and years?

Shortly after the first shots, Palchinsky brought the ministers a
fragment of shell. Admiral Verderevsky recognised the shell as his
own—from a naval gun, from the *Aurora*. But they were shooting
blank from the cruiser. It had been thus agreed, was thus testified by
Flerovsky, and thus reported to the Congress of Soviets later by a
sailor. Was the admiral mistaken? Was the sailor mistaken? Who can
ascertain the truth about a cannon shot fired in the thick of night
from a mutinous ship at a czar's palace where the last government of
the possessing classes is going out like an oilless lamp.

—MAX EASTMAN
(translator)

DOUGLAS MACARTHUR:

FROM

Report of the Chief of Staff

*Douglas MacArthur, son of the founder of the Philippine
Republic, was born in 1880. After graduating from West
Point he engaged in military service in the Far East and
took part in the Vera Cruz expedition in 1905. He commanded the Rainbow Division on the Western Front in
World War I and became chief of staff of the U.S. Army
from 1930 to 1935 in a period of retrenchment and unrest.
On retirement he was appointed military adviser to the
Philippine army and in 1941 was appointed to command
American and Philippine troops in the Pacific. From Bataan
he succeeded in escaping to Australia and as Supreme Commander of the Allied forces in the Southwest Pacific, was
primarily responsible for the amphibious campaigns which
followed. In 1945 he accepted the surrender of the Japanese in Tokyo Bay and for the next five years supervised
the reconstruction of Japan. In 1950, upon the outbreak of
the Korean War, he was appointed commander of the
United Nations forces. He conducted the amphibious landing at Inchon and stemmed the Chinese intervention before
being dismissed by President Truman, in 1951, for his advocacy of the bombing and blockade of China. He died in
1964.*

*"I prepared it," he wrote to B. H. Liddell Hart from Ma-
nila in 1935, "as a doctrinal statement for study by officers
of the American Army and as an educational docu-
ment . . . there is no inflexible pattern upon which to base
military organization even though the objectives are always
firepower and mobility. Every Government has peculiar
factors of geography, climate, finance, industry, and na-*

tional temperament to consider and a particular type of mobility required by one set of circumstances is not necessarily applicable to any other. . . ."

Annual Report of the Chief of Staff for the Fiscal Year June 30, 1935.

To maintain in peace a needlessly elaborate military establishment entails economic waste. But there can be no compromise with minimum requirements. In war there is no intermediate measure of success. Second best is to be defeated, and military defeat carries with it national disaster—political, economic, social, and spiritual disaster. Under the several headings of this report there is sketched in rough outline a military establishment reasonably capable of assuring successful defense of the United States. I have this confidence in its ability, although in size the proposed army is not remotely comparable to many now existing and even falls far below the legal limits prescribed in the National Defense Act.

There are, of course, certain favorable factors which minimize the need in our country for maintenance in peace of a huge military machine such as exists in almost every other major power. Chief among these factors are geographical isolation and the existence of cordial relationships across our land frontiers.

Additionally influencing the determination of the War Department to emphasize quality rather than quantity in further development of the Army is the conviction that relatively small forces exploiting the possibilities of modern weapons and mechanisms will afford in future emergencies a more dependable assurance of defense than will huge, unwieldy, poorly equipped, and hastily trained masses. Adherence to such a policy likewise serves the interests of economy, since of all costs of war, both direct and indirect, none is so irreparable and so devastating as that measured in the blood of its youth.

The United States should not hold to the "nation in arms" as a principal tenet in its doctrine of defense if by that term is indicated an unreasoned purpose of cramming into the armed forces every citizen of military age and capable of carrying a gun. Beyond all doubt any major war of the future will see every belligerent nation highly organized for the single purpose of victory, the attainment of which will require integration and intensification of individual and collective effort. But it will be a nation at war, rather than a nation in arms. Of this vast machine the fighting forces will be only the cutting edge; their mandatory characteristics will be speed in movement, power in fire and shock action, and the utmost in professional skill and leadership. Their armaments will necessarily be of the most efficient types obtainable and the transportation, supply, and maintenance systems supporting them will be required to function perfectly and continuously. Economic and industrial resources will have to assure the adequacy of munition supply and the sustenance of the whole civil population. In these latter fields the great proportion of the employable population will find its war duty.

More than most professions the military is forced to depend upon intelligent interpretation of the past for signposts charting the future. Devoid of opportunity, in peace, for self-instruction through actual practice of his profession, the soldier makes maximum use of historical record in assuring the readiness of himself and his command to function efficiently in emergency. The facts derived from historical analysis he applies to conditions of the present and the proximate future, thus developing a synthesis of appropriate method, organization, and doctrine.

But the military student does not seek to learn from history the minutia of method and technique. In every age these are decisively influenced by the characteristics of weapons currently available and by the means at hand for maneuvering, supplying, and controlling combat forces. But research does bring to light those fundamental principles, and their combinations and applications, which, in the past, have been productive of success. These principles know no limitation of time. Consequently, the army extends its analytical interest to the dust-buried accounts of wars long past as well as to those still reeking with the scent of battle. It is the object of the search that dictates the field for its pursuit. Those callow critics who hold that only in the most recent battles are there to be found truths applicable to our present problems have failed utterly to see this. They apparently cling to a fatuous hope that in historical study is to be found a complete digest of the science of war rather than simply the basic and inviolable laws of the art of war.

Were the accounts of all battles, save only those of Genghis

Khan, effaced from the pages of history, and were the facts of his campaigns preserved in descriptive detail, the soldier would still possess a mine of untold wealth from which to extract nuggets of knowledge useful in molding an army for future use. The success of that amazing leader, beside which the triumphs of most other commanders in history pale into insignificance, are proof sufficient of his unerring instinct for the fundamental qualifications of an army.

He devised an organization appropriate to conditions then existing; he raised the discipline and the morale of his troops to a level never known in any other army, unless possibly that of Cromwell; he spent every available period of peace to develop subordinate leaders and to produce perfection of training throughout the army, and, finally, he insisted upon speed in action, a speed which by comparison with other forces of his day was almost unbelievable. Though he armed his men with the best equipment of offense and of defense that the skill of Asia could produce, he refused to encumber them with loads that would immobilize his army. Over great distances his legions moved so rapidly and secretly as to astound his enemies and practically to paralyze their powers of resistance. He crossed great rivers and mountain ranges, he reduced walled cities in his path and swept onward to destroy nations and pulverize whole civilizations. On the battlefield his troops maneuvered so swiftly and skillfully and struck with such devastating speed that times without number they defeated armies overwhelmingly superior to themselves in numbers.

Regardless of his destructiveness, his cruelty, his savagery, he clearly understood the unvarying necessities of war. It is these conceptions that the modern soldier seeks to separate from the details of the Khan's technique, tactics, and organization, as well as from the ghastly practices of his butcheries, his barbarism, his ruthlessness. So winnowed from the chaff of medieval custom and of all other inconsequentials, they stand revealed as kernels of eternal truth, as applicable today in our efforts to produce an efficient army as they were when, seven centuries ago, the great Mongol applied them to the discomfiture and amazement of a terrified world. We cannot violate these laws and still produce and sustain the kind of army that alone can insure the integrity of our country and the permanency of our institutions if ever again we face the grim realities of war.

T. E. LAWRENCE:

FROM

The Evolution of a Revolt;

FROM

The Seven Pillars of Wisdom

T. E. (Thomas Edward) Lawrence was born in Wales in 1885, the illegitimate son of an Irish baronet. After graduating from Oxford with a degree in modern history and undertaking archaeological work in the Middle East, he joined the army intelligence staff at Cairo, at the outbreak of war with Turkey. In 1916 he was sent on a mission to the Hejaz where the Arabs had started a revolt against the Turks.

Attached to the staff of the principal Arab leader, Feisal (largely selected by Lawrence and later the ruler of Iraq), he administered British support for the Arab revolt and largely directed its strategy in the field. This led to his occupation of Damascus in 1918 and was, in large measure, instrumental in the collapse of the Ottoman Empire. He also created a legend. He attended the peace conference in 1919 but, frustrated by its outcome as far as Arab expectations were concerned, he devoted himself to writing The Seven Pillars of Wisdom.

He had been promoted colonel in World War I and was offered high posts afterwards, but in 1922 he enlisted in the ranks of the recently created Royal Air Force. He transferred to the tank corps and then reenlisted in the air force, where he remained in the ranks until his retirement in 1935, working on the development of power boats. He was killed riding a motor bike in the same year. The Mint, *his book about barrack-room life, was published posthumously in 1955.*

FROM *The Evolution of a Revolt*

My own personal duty was to command, and I began to unravel command and analyse it, both from the point of view of strategy, the aim in war, the synoptic regard which sees everything by the standard of the whole, and from the point of view called tactics, the means towards the strategic end, the steps of its staircase.

In each I found the same elements, one algebraical, one biological, a third psychological. The first seemed a pure science, subject to the laws of mathematics, without humanity. It dealt with known invariables, fixed conditions, space and time, inorganic things like hills and climates and railways, with mankind in type—masses too great for individual variety, with all artificial aids, and the extensions given to our faculties by mechanical invention. It was essentially formulable. . . .

The second factor was biological, the breaking-point, life and death, or better, wear and tear. Bionomics seemed a good name for it. The war-philosophers had properly made it an art, and had elevated one item in it, "effusion of blood," to the height of a principle. It became humanity in battle, an art touching every side of our corporal being, and very warm. There was a line of variability (man) running through all its estimates. Its components were sensitive and illogical, and generals guarded themselves by the device of a reserve, the significant medium of their art. . . .

Nine-tenths of tactics are certain and taught in books: but the irrational tenth is like the kingfisher flashing across the pool, and that is the test of generals. It can only be ensued by instinct, sharpened by thought practising the stroke so often that at the crisis it is as natural as a reflex. . . .

The third factor in command seemed to be psychological, that science (Xenophon called it diathetic) of which our propaganda is a stained and ignoble part. . . . The printing press is the greatest weapon in the armoury of the modern commander, and we, being amateurs in the art of command, began our war in the atmosphere of the twentieth century, and thought of our weapons without prejudice, not distinguishing one from another socially. The regular officer has the tradition of forty generations of serving soldiers behind him, and to him the old weapons are the most honoured. We had seldom to concern ourselves with what our men did, but much with what they thought, and to us the diathetic was more than half command. In Europe it was set a little aside and entrusted to men outside the General Staff. In Asia we were so weak physically that we could not let the metaphysical weapon rust unused. We had won a province when we had taught the civilians in it to die for our ideal of freedom: the presence or absence of the enemy was a secondary matter. . . .

Napoleon had said it was rare to find generals willing to fight battles. The curse of this war was that so few could do anything else. Napoleon had spoken in angry reaction against the excessive finesse of the eighteenth century, when men almost forgot that war gave them license to murder. We had been swinging out on his dictum for a hundred years and it was time to get back a bit again. . . . Our cards were speed and time, not hitting power, and these gave us strategical rather than tactical strength. Range is more to strategy than force. The invention of bully-beef has modified land-war more profoundly than the invention of gun-powder.

My chiefs did not follow all these arguments, but gave me leave to try my hand after my own fashion. We went off first to Akaba, and took it easily. Then we took Tafileh and the Dead Sea: then Azrak and Deraa, and finally Damascus, all in successive stages worked out consciously on these sick-bed theories. . . .

In character these operations were more like warfare than ordinary land operations, in their mobility, their ubiquity, their independence of bases and communications, their lack of ground features, of strategic areas, of fixed directions, of fixed points. "He who commands the sea is at great liberty, and may take as much or as little of the war as he will": he who commands the desert is equally fortunate.

FROM *The Seven Pillars of Wisdom*

This last straight bank, with Byzantine foundations in it, seemed very proper for a reserve or ultimate line of defence for Tafileh. To be sure, we had no reserve as yet—no one had the least notion who or what we would have anywhere—but, if we did have anybody, here was their place: and at that precise moment Zeid's personal Ageyl became visible, hiding coyly in a hollow. To make them move required words of a strength to unravel their plaited hair: but at last I had them sitting along the skyline of Reserve Ridge. They were about twenty, and from a distance looked beautiful, like "points" of a considerable army. I gave them my signet as a token, with orders to collect there all newcomers, especially my fellows with their gun.

As I walked northward towards the fighting, Abdulla met me, on his way to Zeid with news. He had finished his ammunition, lost five men from shell-fire, and had one automatic gun destroyed. Two guns, he thought the Turks had. His idea was to get up Zeid with all his men and fight: so nothing remained for me to add to his message; and there was no subtlety in leaving alone my happy masters to cross and dot their own right decision.

He gave me leisure in which to study the coming battlefield. The tiny plain was about two miles across, bounded by low green ridges, and roughly triangular, with my reserve ridge as base. Through it ran the road to Kerak, dipping into the Hesa valley. The Turks were fighting their way up this road. Abdulla's charge had taken the western or left-hand ridge, which was now our firing-line.

Shells were falling in the plain as I walked across it, with harsh stalks of wormwood stabbing into my wounded feet. The enemy fusing was too long, so that the shells grazed the ridge and burst away behind. One fell near me, and I learned its calibre from the hot cap. As I went they began to shorten range, and by the time I got to the ridge it was being freely sprinkled with shrapnel. Obviously the Turks had got observation somehow, and looking round I saw them climbing along the eastern side beyond the gap of the Kerak road. They would soon outflank us at our end of the western ridge.

"Us" proved to be about sixty men, clustered behind the ridge in two bunches, one near the bottom, one by the top. The lower was made up of peasants, on foot, blown, miserable, and yet the only warm things I had seen that day. They said their ammunition was finished, and it was all over. I assured them it was just beginning and pointed to my populous reserve ridge, saying that all arms were there in support. I told them to hurry back, refill their belts and hold on to it for good. Meanwhile we would cover their retreat by sticking here for the few minutes yet possible.

They ran off, cheered, and I walked about among the upper group quoting how one should not quit firing from one position till ready to fire from the next. In command was young Metaab, stripped to his skimp riding-drawers for hard work, with his black love-curls awry, his face stained and haggard. He was beating his hands together and crying hoarsely with baffled vexation, for he had meant to do so well in this, his first fight for us.

My presence at the last moment, when the Turks were breaking through, was bitter; and he got angrier when I said that I only wanted to study the landscape. He thought it flippancy, and screamed something about a Christian going into battle unarmed. I retorted with a quip from Clausewitz, about a rearguard effecting its purpose more by being than by doing: but he was past laughter, and perhaps with justice, for the little flinty bank behind which we sheltered was crackling with fire. The Turks, knowing we were there, had turned twenty machine-guns upon it. It was four feet high and fifty feet long, of bare flinty ribs, off which the bullets slapped deafeningly: while the air above so hummed or whistled with ricochets and chips that it felt like death to look over. Clearly we must leave very soon, and as I had no horse I went off first, with Metaab's promise that he would wait where he was if he dared, for another ten minutes.

The run warmed me. I counted my paces, to help in ranging the Turks when they ousted us; since there was only that one position for them, and it was poorly protected against the south. In losing this Motalga ridge we would probably win the battle. The horsemen held on for almost their ten minutes, and then galloped off without hurt. Metaab lent me his stirrup to hurry me along, till we found ourselves breathless among the Ageyl. It was just noon, and we had leisure and quiet in which to think.

Our new ridge was about forty feet up, and a nice shape for defence. We had eighty men on it, and more were constantly arriving. My guards were in place with their gun; Lutfi, an engine-destroyer, rushed up hotly with his two, and after him came another hundred Ageyl. The thing was becoming a picnic, and by saying "excellent" and looking overjoyed, we puzzled the men, and made them consider the position dispassionately. The automatics were put on the skyline, with orders to fire occasional shots, short, to disturb the Turks a little, but not too much, after the expedient of Massena in delaying enemy deployment. Otherwise a lull fell; I lay down in a sheltered place which caught a little sun, and no wind, and slept a blessed hour, while the Turks occupied the old ridge, extending over it like a school of geese, and about as wisely. Our men left them alone, being contented with a free exhibition of themselves.

In the middle of the afternoon Zeid arrived, with Mastur, Rasim and Abdulla. They brought our main body, comprising twenty mounted infantry on mules, thirty Motalga horsemen, two hundred villagers, five automatic rifles, four machine-guns and the Egyptian Army mountain gun which had fought about Medina, Petra and Jurf. This was magnificent, and I woke up to welcome them.

The Turks saw us crowding, and opened with shrapnel and machine-gun fire: but they had not the range and fumbled it. We reminded one another that movement was the law of strategy, and started moving. Rasim became a cavalry officer, and mounted with all our eighty riders of animals to make a circuit about the eastern ridge and envelop the enemy's left wing, since the books advised attack not upon a line, but upon a point, and by going far enough along any finite wing it would be found eventually reduced to a point of one single man. Rasim liked this, my conception of his target.

He promised, grinningly, to bring us that last man: but Hamd el Arar took the occasion more fittingly. Before riding off he devoted himself to the death for the Arab cause, drew his sword ceremoniously, and made to it, by name, a heroic speech. Rasim took five automatic guns with him; which was good.

We in the centre paraded about, so that their departure might be unseen of the enemy, who were bringing up an apparently endless procession of machine-guns and dressing them by the left at intervals along the ridge as though in a museum. It was lunatic tactics. The ridge was flint, without cover for a lizard. We had seen how, when a bullet struck the ground, it and the ground spattered up a shower of deadly chips. Also we knew the range and elevated our Vickers guns carefully, blessing their long, old-fashioned sights; our mountain gun was propped into place ready to let go a sudden burst of shrapnel over the enemy when Rasim was at grips.

As we waited, a reinforcement was announced of one hundred men from Aima. They had fallen out with Zeid over war-wages the day previous, but had grandly decided to sink old scores in the crisis. Their arrival convinced us to abandon Marshal Foch and to attack from, at any rate, three sides at once. So we sent the Aima men, with three automatic guns, to outflank the right, or western wing. Then we opened against the Turks from our central position, and bothered their exposed lines with hits and ricochets.

The enemy felt the day no longer favourable. It was passing, and sunset often gave victory to defenders yet in place. Old General Hamid Fakhri collected his Staff and Headquarters, and told each man to take a rifle. "I have been forty years a soldier, but never saw I rebels fight like these." Enter the ranks . . . but he as too late. Rasim pushed forward an attack of his five automatic guns, each with its two-man crew. They went in rapidly, unseen till they were in position, and crumpled the Turkish left.

The Aima men, who knew every blade of grass on these, their own village pastures, crept, unharmed, within three hundred yards of the Turkish machine-guns. The enemy, held by our frontal threat, first knew of the Aima men when they, by a sudden burst of fire, wiped out the gun-teams and flung the right wing into disorder. We saw it, and cried advance to the camel men and levies about us.

Mohamed el Ghasib, comptroller of Zeid's household, led them on his camel, in shining wind-billowed robes, with the crimson banner of the Ageyl over his head. All who had remained in the centre with us, our servants, gunners and machine-gunners, rushed after him in a wide vivid line.

The day had been too long for me, and I was now only shaking with desire to see the end: but Zeid beside me clapped his hands with joy at the beautiful order of our plan unrolling in the frosty redness of the setting sun. On the one hand Rasim's cavalry were sweeping a broken left wing into the pit beyond the ridge: on the other the men of Aima were bloodily cutting down fugitives. The enemy centre was pouring back in disorder through the gap, with our men after them on foot, on horse, on camel. The Armenians, crouching behind us all day anxiously, now drew their knives and howled to one another in Turkish as they leaped forward.

I thought of the depths between here and Kerak, the ravine of

Hesa, with its broken, precipitous paths, the undergrowth, the narrows and defiles of the way. It was going to be a massacre and I should have been crying-sorry for the enemy; but after the angers and exertions of the battle my mind was too tired to care to go down into that awful place and spend the night saving them. By my decision to fight, I had killed twenty or thirty of our six hundred men, and the wounded would be perhaps three times as many. It was one-sixth of our force gone on a verbal triumph, for the destruction of this thousand poor Turks would not affect the issue of the war.

In the end we had taken their two mountain howitzers (Skoda guns, very useful to us), twenty-seven machine-guns, two hundred horses and mules, two hundred and fifty prisoners. Men said only fifty got back, exhausted fugitives, to the railway. The Arabs on their track rose against them and shot them ignobly as they ran. Our own men gave up the pursuit quickly, for they were tired and sore and hungry, and it was pitifully cold. A battle might be thrilling at the moment for generals, but usually their imagination played too vividly beforehand, and made the reality seem sham; so quiet and unimportant that they ranged about looking for its fancied core. This evening there was no glory left, but the terror of the broken flesh, which had been our own men, carried past us to their homes.

ADOLF HITLER:

FROM

My Struggle (Mein Kampf)

Adolf Hitler was born in Austria in 1889, the son of a customs official, and after leaving school at sixteen engaged in minor artistic pursuits. At the outbreak of World War I he volunteered for the German army, was promoted corporal in an infantry regiment and was gassed. He became an army political agent at the end of the war and from 1920 on devoted himself to the creation of the National Socialist party. An unsuccessful attempt to seize power with Ludendorff in 1923 was followed by five years' imprisonment. In 1932 he was constitutionally appointed chancellor and quickly gained absolute power. After defeat by the Allies in World War II he committed suicide in his Berlin bunker in 1945.

My Struggle *was first published in 1925.*

Hence Germany's only hope of carrying out a sound territorial policy lay in acquiring fresh lands in Europe itself. Colonies are useless for that object if they appear unsuitable for settling Europeans in large numbers. In the Nineteenth Century, however, it was no longer possible to acquire such territory for colonization by peaceful methods. A colonizing policy of that kind could only be realized by means of a hard struggle, which would be far more appropriate for the sake of gaining territory in the continent near home than for lands outside Europe.

For such a policy there was only one possible ally in Europe—Great Britain. Great Britain was the only Power which could protect our rear, supposing we started a new Germanic expansion. We should have had as much right to do this as our forefathers had.

No sacrifice would have been too great in order to gain England's alliance. It would have meant renunciation of colonies and importance on the sea, and refraining from interference with British industry by our competition.

There was a moment when Great Britain would have let us speak to her in this sense; for she understood very well that, owing to her increased population, Germany would have to look for some solution and find it either in Europe with Great Britain's help, or elsewhere in the world without it.

The attempt made from London at the turn of the century to obtain a rapprochement with Germany was due first and foremost to this feeling. But the Germans were upset by the idea of "having to pull England's chestnuts out of the fire for her"—as if an alliance were possible on any basis other than that of reciprocity. On that principle business could very well have been done with Whitehall. British diplomacy was quite clever enough to know that nothing could be hoped for without reciprocity.

Let us imagine that Germany, with a skilful foreign policy, had played the part which Japan played in 1904—we can hardly estimate the consequences that would have had for Germany.

There would never have been a World War.

That method, however, was never adopted at all.

There still remained the possibility: industry and world trade, sea power and colonies.

If a policy of territorial acquisition in Europe could only be pursued in alliance with Great Britain against Russia, a policy of colonies and world trade, on the other hand, was only conceivable in alliance with Russia against Great Britain. In this case they should have drawn their conclusion ruthlessly, and have sent Austria packing.

They adopted a formula of "peaceful economic conquest of the world," which was destined to destroy for ever the policy of force which they had pursued up to that time. Perhaps they were not quite certain of themselves at times when quiet incomprehensible threats came across from Great Britain. Finally they made up their minds to build a fleet, not for the purpose of attacking and destroying, but to defend the "world-peace" and for the "peaceful conquest of the world." Thus they were constrained to maintain it on a modest scale, not only as regards numbers, but also as regards the tonnage of individual ships and their armaments, so as to make it evident that their final aim was a peaceful one.

The talk about "peaceful economic conquest of the world" was the greatest piece of folly ever set up as a leading principle in State policy, especially as they did not shrink from quoting Britain to prove that it was possible to carry it out in practice. The harm done by our professors with their historical teaching and theories can scarcely be made good again, and it merely proves in a striking fashion how many "learn" history without understanding it or taking it in. Even in the British Isles they had had to confess to a striking refutation of the theory; and yet no nation ever prepared better for economic conquest even with the sword, or later maintained it more ruthlessly, than the British.

Is it not the hallmark of British statecraft to make economic gains out of political strength and at once to reconvert each economic gain into political power? Thus it was a complete error to imagine that England personally was too cowardly to shed her blood in defence of her economic policy! The fact that the British possessed no national army was no proof to the contrary; for it is not the military form of the national forces that matters, but rather the will and determination to make use of what there is. England always possessed the armaments which she needed. She always fought with whatever weapons were necessary to ensure success. She fought with mercenaries as long as mercenaries were good enough; but she seized hold of the best blood in all the nation whenever such a sacrifice was needed to make victory sure, and she had always determination to fight, and was tenacious and unflinching in the conduct of her wars.

In Germany, however, as time went on they encouraged, by means of the schools, the Press and the comic papers, an idea of British life and even more so, of the Empire, which was bound to lead to the most ill-timed self-deception; for everything became gradually contaminated with this rubbish, and the result was a low opinion of the British, which ended by revenging itself most bitterly. This mistaken idea ran so deeply that everyone was convinced that the Englishman, as they imagined him, was a business man, both crafty and incredibly cowardly. It never occurred to our worthy professorial imparters of knowledge that anything as vast as the British world Empire could never have been assembled and kept together merely by swindling and underhand methods. The few who gave warnings were either ignored or silenced. I remember distinctly the amazement on the faces of my comrades in arms when we came face to face with the Tommies in Flanders. After the very first days of fighting it dawned on the brain of each man that those Scotchmen did not exactly correspond with the people whom writers in comic papers and newspaper reports had thought fit to describe to us.

I began to reflect then on propaganda and the most useful forms of it.

—JAMES MURPHY
(translator)

ALBERT SPEER:

FROM

Inside the Third Reich

*Albert Speer was born in Mannheim in 1905. He grew up
in liberal and artistic surroundings and graduated in architecture. In 1932 he was asked by Hitler to become his
private architect. After the Nazis came to power he designed their great public works, including the Reich Chan-
cellory in Berlin. Though he had previously been scarcely
involved in party activities, he became one of the closest
personal associates of Hitler.*

*He was appointed armaments minister in 1942 with
wide powers over industrial production and labor throughout Germany and occupied Europe. In the end he plotted
to kill Hitler and, in 1945, successfully defied his final order
to carry out a scorched earth policy before the Allied advance. At the Nuremburg trial of the major war criminals
he was sentenced to twenty years imprisonment for crimes
against humanity. He was released from Spandau Prison
in 1966.*

The victories of the early years of the war can literally be attributed to Hitler's ignorance of the rules of the game and his lay-
man's delight in decision-making. Since the opposing side was trained
to apply rules which Hitler's self-taught, autocratic mind did not
know and did not use, he achieved surprises. These audacities,
coupled with military superiority, were the basis of his early
successes. But as soon as setbacks occurred he suffered shipwreck,
like most untrained people. Then his ignorance of the rules of the
game was revealed as another kind of incompetence; then his defects

were no longer strengths. The greater the failures became, the more obstinately his incurable amateurishness came to the fore. The tendency to wild decisions had long been his forte; now it speeded his downfall.

Every two or three weeks I travelled from Berlin to spend a few days in Hitler's East Prussian, and later in his Ukrainian, Headquarters in order to have him decide the many technical questions of detail in which he was interested in his capacity as Commander-in-Chief of the army. Hitler knew all the types of ordnance and ammunition, including the calibers, the lengths of barrels, and the range of fire. . . . The real expert does not burden his mind with details that he can look up or leave to an assistant. Hitler, however, felt it necessary for his own self-esteem to parade his knowledge. But he also enjoyed doing it. He obtained his information from a large book in a red binding with broad yellow diagonal stripes. It was a catalogue being brought up-to-date, of from thirty to fifty different types of ammunition and ordnance. He kept it on his night table. Sometimes he would order a servant to bring the book down when in the course of military conferences a colleague had mentioned a figure which Hitler instantly corrected. The book was opened, Hitler's data would be confirmed, without fail, every time, while the General would be shown to be in error. Hitler's memory for figures was the terror of his entourage.

It often seemed to me that Hitler used these prolonged conferences on armaments and war production as an escape from his military responsibilities. He himself admitted to me that he found in them a relaxation similar to our former conferences on architecture. Even in a crisis situation he devoted many hours to such discussions, sometimes refusing to interrupt them even when his Field Marshals or ministers urgently wanted to speak to him.

Our technical conferences were usually combined with a demonstration of new weapons which took place in a nearby field. . . . Often Hitler and I would make appreciative remarks about the weapons, such as "What an elegant barrel" or "What a fine shape this tank has!"—a ludicrous relapse into the terminology of our joint inspections of architectural models.

In the course of one such inspection, Keitel mistook a 7.5 centimeter anti-tank gun for a light field howitzer. Hitler passed over the mistake at the time but had his joke on our ride back. "Did you hear that? Keitel and the anti-tank gun? And he's a general of the artillery!"

* * *

During the next twenty years of my life I was guarded in Spandau prison, by nations of the Four Powers against whom I had organized Hitler's war. Along with my six fellow-prisoners, they were the only people I had close contact with. Through them I learned directly what the effects of my work had been. Many of them mourned loved ones who had died in the war—in particular, every one of the Soviet guards had lost some close relative, brothers or a father. Yet not one of them bore a grudge towards me for my personal share in the tragedy; never did I hear words of recrimination. At the lowest ebb of my existence, in contact with these ordinary people, I encountered uncorrupted feelings of sympathy, helpfulness, human understanding, feelings that by-passed the prison rules. . . . And now at last I wanted to understand.

—RICHARD AND CLARA WINSTON
(translators)

CHARLES DE GAULLE:

FROM

The Edge of the Sword

Charles de Gaulle was born in 1890, and after attending St. Cyr, joined an infantry regiment under the command of Philippe Pétain. He was taken prisoner at Verdun in 1916 and became a lecturer at St. Cyr after the Armistice. His views aroused considerable interest and some official opposition. At the outbreak of World War II he commanded a tank brigade and then an armored division on the Western Front. After the German breakthrough he was appointed Under-Secretary of War and after the Armistice put himself at the head of the Free French forces in London. He formed a provisional government after the liberation but resigned in 1946 in a disagreement over the constitutional arrangements for the Fourth Republic. He eventually returned to power in 1958 during the Algerian crisis, suppressed the generals' revolt and granted Algeria independence. As first president of the Fifth Republic he embarked on a program of domestic reform and military strength, including nuclear weapons. He resigned in 1969 and died in 1971.

The Edge of the Sword *was published prophetically in 1932.*

Great war-leaders have always been aware of the importance of instinct. Was not what Alexander called his "hope," Caesar his "luck" and Napoleon his "star" simply the fact that they knew they had a particular gift of making contact with realities sufficiently closely to dominate them? For those who are greatly gifted, this faculty often shines through their personalities. There may be nothing in itself exceptional about what they say or their way of saying it, but other men in their presence have the impression of a natural force destined to master events. Flaubert expresses this feeling when he describes the still adolescent Hannibal as already clothed "in the undefinable splendour of those who are destined for great enterprises."

However, while no work or action can be conceived without the promptings of instinct, these promptings are not sufficient to give conception a precise form. The very fact that they are "gifts of nature" means that they are simple, crude and sometimes confused. Now, a general commands an army, that is to say a system of complex forces with its own properties and disciplines, which can develop its power only by following a certain pattern. It is here that the intelligence comes into its own. Taking possession of the raw material of instinct, it elaborates it, gives it a specific shape, and makes of it a clearly defined and coherent whole. . . .

If a commander is to grasp the essentials and reject the inessentials; if he is to split his general operation into a number of complementary actions in such a way that all shall combine to achieve the purpose common to every one of them, he must be able to see the situation as a whole, to attribute to each object its relative importance, to grasp the connections between each factor in the situation and to recognize its limits. All this implies a gift of synthesis which, in itself, demands a high degree of intellectual capacity. The general who has to free the essentials of his problem from the confused mass of their attendant details, resembles the user of a stereoscope who has to concentrate his eyes upon the image before he can see it in relief. That is why great men of action have always been of the meditative type. They have, without exception, possessed, to a very high degree, the faculty of withdrawing into themselves. As Napoleon said: "The military leader must be capable of giving intense, extended and indefatigable consideration to a single group of objects."

If the conceiving of an action is to be valid, which means adapted to the circumstances of the case, it calls for a combined effort of intelligence and instinct. Critics of action in warfare have, however, rarely been willing to admit that these two faculties have each a necessary part to play, though no one of them is able to do without the other. Often, the critic has seen fit arbitrarily to break the balance between the two faculties and has attributed to one of them alone the whole responsibility for having produced the concept of action in question.

Certain critics, recognizing the powerlessness of reason alone to solve the many problems involved, have gone so far as to maintain that it is impossible, in war, for any leader to dominate events, since, no matter how great his intellectual gifts may be, he cannot control the action itself. "There is no such thing"—they say—"as an art of war, since, in the last analysis, it is chance alone, that determines the outcome of battles." Philosophers and writers are only too willing to adopt this sceptical attitude, and this is not hard to explain. Minds exclusively devoted to speculation lose the sense of what action requires. Armed with the one instrument they are familiar with, that of the pure intelligence, they fail to penetrate into the inner sense of the action, and convert their failure to comprehend into disdain.

Thus, Socrates, when engaged in argument with Nicomachides who was complaining that the Assembly had appointed an incompetent citizen to be their general, maintained that it did not matter at all, since events would have taken the same course even if some able and conscientious commander had been chosen. It is true, however, that the same Socrates, when questioned by Pericles about the causes of indiscipline among the Athenian troops, placed the responsibility for this state of affairs on the officers unfit to exercise command.

A similar attitude led Tolstoy, in *War and Peace,* to describe Bagration at Hollabrunn as letting events, which he thought he could not change, take their course, and confining his efforts "to making what was the result of chance, *look* as though everything had happened in accordance with his orders or his intentions."

So, too, Anatole France makes Jérome Coignard say: "When two hostile armies meet, one is bound to be defeated, whence it follows that the other must necessarily be victorious, though its commander lack some, or, indeed, all of the qualities that make a great leader. How, then, is it possible"—concludes the philosophical abbé—"to distinguish in such combats what is the effect of art from what is the gift of fortune?"

Nor must we forget King Ubu who won a victory just *because* he had taken no preliminary steps of any kind.

It is true that military men, exaggerating the relative powerlessness of the intelligence, will sometimes neglect to make use of it at all. Here the line of least resistance comes into its own. There have been examples of commanders avoiding all intellectual effort and even despising it on principle. Every great victory is usually followed by this kind of mental decline. The Prussian Army after the death of Frederick the Great is an instance of this. In other cases, the military men note the inadequacy of knowledge and therefore trust to inspiration alone or to the dictates of fate. That was the prevalent state of mind of the French Army at the time of the Second Empire: "We shall muddle through, somehow."

Often, on the other hand, the intellect is unwilling to allow instinct its proper share. Absolute master in the field of speculation, intelligence refuses to share the empire of action and attempts to impose itself alone. When this happens, the true nature of war is completely misunderstood, and those responsible for its conduct try to apply to it a rigid and therefore arbitrary set of rules.

Accustomed to working from "solid" premises, the unaided intelligence wants to deduce its conception from constants known in advance, whereas what is needed is to induce the conception from contingent and variable facts in each individual case.

This tendency, it is true, exercises a special attraction over the French mind. Inquisitive and quick in the uptake, the Frenchman feels the need for logic, likes linking a series of facts by a process of reasoning, and trusts more readily to theory than to experience.

This natural slant is accentuated by the categorical nature of military discipline and reinforced by the dogmatism inherent in education. Consequently "schools of thought" flourish in France more than in any other country. Their absolute and speculative character render these schools of thought attractive and dangerous, and they have already cost us dear.

—GERARD HOPKINS
(translator)

MAO TSE-TUNG:

FROM

On Guerrilla Warfare;

FROM

Anti-Japanese Guerrilla Warfare

Mao Tse-tung was born in central China in 1893, of a peasant family, and after educating himself became a librarian at Peking University. He helped to found the Chinese Communist party in 1921 and promoted several rebellions against the central government. In 1931, with Russian support, he established a soviet republic in the remote interior. Overrun by government forces, he led his followers on the "Long March" to a new base in the northwest and then entered into an alliance against the Japanese invader. After the Japanese surrender he turned against Chiang Kai-shek and by 1949 had gained complete control of the mainland. He proclaimed a People's Republic and remained in power.

FROM *On Guerrilla Warfare*, 1936

The collections of military rules and orders promulgated in many countries point out the necessity of "applying principles elastically according to the situation," as well as the measures to be taken in a defeat. The former requires a commander not to commit mistakes subjectively through too flexible an application of principles, while the latter tells him how to cope with a situation when he has already committed mistakes or when unexpected and irresistible changes occur in the circumstances.

Why are mistakes committed? Because the disposition of forces in the war or battle or the directing of them does not fit in with the conditions of a certain time and a certain place, because the directing does not correspond with or dovetail into realities, in other words, because the contradiction between the subjective and the objective is not solved. People can hardly avoid coming up against such a situation in performing any task, only some are more and others are less competent in performing it. We demand greater competence in performing tasks, and in war we demand more victories or, conversely, fewer defeats. The crux here lies precisely in making the subjective and the objective correspond well with each other. . . .

The process of knowing the situation goes on not only before but also after the formulation of a military plan. The carrying out of a plan, from its very beginning to the conclusion of an operation, is another process of knowing the situation, *i.e.* the process of putting it into practice. In this process, there is need to examine anew whether the plan mapped out in the earlier process corresponds with the actualities. If the plan does not correspond or does not fully correspond with them, then we must, according to fresh knowledge, form new judgments and make new decisions to modify the original plan in order to meet the new situation. There are partial modifications in almost every operation, and sometimes even a complete change. A hothead who does not know how to change his plan, or is unwilling to change it but acts blindly, will inevitably run his head against a brick wall.

The above applies to a strategical action, a campaign, or a battle. If an experienced military man is modest and willing to learn, and has familiarised himself with the conditions of his own forces (officers and men, arms, supplies, etc., and their totality) as well as those of the enemy (similarly, officers and men, arms, supplies, etc., and their totality), and with all other conditions relating to war, such as politics, economy, geography and weather conditions, he will be more confident in directing a war or an operation and will be more likely to win it. This is because over a long period of time he has learnt the situation on both the enemy side and his own, discovered the laws of action, and solved the contradiction between the subjective and the objective. This process of knowing is very important; without such a long period of experience it is difficult to understand

and grasp the laws of an entire war. No truly able commander of a high rank can be made out of one who is a mere beginner in warfare or one who knows warfare only on paper; and to become such a commander one must learn through warfare.

All military laws and theories partaking of the character of principle represent past military experiences summed up by people in both ancient and modern times. We should carefully study the lessons which were learnt in past wars at the cost of blood and which have been bequeathed to us. This is one point. But there is another point, namely, we must also put conclusions thus reached to the test of our own experience and absorb what is useful, reject what is useless, and add what is specifically our own. The latter is a very important point, for otherwise we cannot direct a war.

Reading books is learning, but application is also learning and the more important form of learning. To learn warfare through warfare—this is our chief method. A person who has had no opportunity to go to school can also learn warfare, which means learning it through warfare. As a revolutionary war is the concern of the masses of the people, it is often undertaken without previous learning but is learnt through undertaking it—undertaking is itself learning. There is a distance between a civilian and a soldier, but that distance is not as long as the Great Wall and can be quickly eliminated; to take part in revolution and war is the method of eliminating it. To say that learning and application are difficult means that it is difficult to learn thoroughly and apply skilfully. To say that civilians can very quickly become soldiers means that it is not difficult to get them initiated. In summarising these two aspects we may apply an old Chinese adage: "Nothing is difficult in the world for anyone who sets his mind on it." Initiation is not difficult and mastery is also possible so long as one sets one's mind on them and is good at learning.

Military laws, like the laws governing all other things, are a reflection in our mind of objective realities; everything is objective reality except our mind. Consequently what we want to learn and know includes things both on the enemy side and our own, and both sides should be regarded as the object of our study and only our mind (thinking capacity) is the subject that makes the study. Some people are intelligent in knowing themselves but stupid in knowing their opponents, and others are the other way round; neither kind can solve the problem of learning and applying the laws of war. We must not belittle the saying in the book of Sun Tzu, the great military expert of ancient China, "Know your enemy and know yourself, and you can fight a hundred battles without disaster," a saying which refers both to the stage of learning and to the stage of application, both to knowing laws of the development of objective realities and to deciding on our own action according to them in order to overcome the enemy facing us.

War is the highest form of struggle between nations, states, classes, or political groups, and all laws of war are applied by a nation, a state, a class, or a political group waging a war to win victory for itself. It is beyond question that success or failure in a war is mainly determined by the military, political, economic and natural conditions on both sides. But not entirely so; it is also determined by the subjective ability on each side in directing the war. A military expert cannot expect victory in war by going beyond the limits imposed by material conditions, but within these limits he can and must fight to win. The stage of action of a military expert is built upon objective material conditions, but with the stage set, he can direct the performance of many lively dramas, full of sound and colour, of power and grandeur.

Swimming in an immense ocean of war, a commander must not only keep himself from sinking but also make sure to reach the opposite shore with measured strokes. The laws of directing wars constitute the art of swimming in the ocean of war.

—ANNE FREEMANTLE
(translator)

FROM *Anti-Japanese Guerrilla Warfare, 1937*

A guerrilla unit should carry out the task of extricating itself from a passive position, when it is forced into one through some incorrect estimation and disposition, or some overwhelming pressure. How to extricate itself from it depends on circumstances. The circumstances are often such as to make it necessary to "run away." The ability to run away is precisely one of the characteristics of the guerrillas. Running away is the chief means of getting out of passivity and regaining the initiative. But it is not the only means. The moment when the enemy exerts maximum pressure and we are in the worst predicament often happens to be the very point at which he begins to be at a disadvantage and we begin to enjoy advantages. Frequently the initiative and an advantageous position are gained through one's effort of "holding out a bit longer."

Now we shall deal with flexibility.

Flexibility is a concrete manifestation of initiative. Flexible employment of forces is more indispensable in guerrilla warfare than in regular warfare.

The directors of guerrilla war must understand that the flexible employment of forces is the most important means of changing the situation between the enemy and ourselves and gaining the initiative. As determined by the special features of guerrilla warfare, guerrilla forces must be flexibly employed according to conditions, such as the task, the enemy disposition, the terrain and the inhabitants; and the chief ways of employing the forces consist in dispersing, concentrating and shifting them.

In employing the guerrilla units, the director of guerrilla war is like a fisherman casting a net which he should be able to spread out as well as to draw in. When a fisherman spreads out his net, he must first find out the depth of the water, the speed of the current and the presence or absence of obstructions, similarly when the guerrilla units are dispersed we must also be careful not to incur losses through an ignorance of the situation and mistakes in actions. A fisherman, in order to draw in his net, must hold fast the end of the cord; in employing the forces, it is also necessary to maintain liaison and communication and to keep an adequate portion of the main force to hand. As a fisherman must frequently change his place, so guerrillas should constantly shift their positions. Dispersion, concentration and shifting of the forces are the three ways of flexibly employing the forces of guerrilla warfare.

Generally speaking, the dispersion of guerrilla units, *i.e.* "breaking up the whole into parts," is employed mainly in the following circumstances: (1) when we threaten the enemy with a wide frontal attack because he is on the defensive and we are still unable to mass our forces to engage him; (2) when we widely harass and disrupt the enemy in an area where his forces are weak; (3) when, unable to break through the enemy's encirclement, we try to disperse his attention in order to get away from him; (4) when we are restricted by the condition of terrain or in matters of supply; or (5) when we carry on work among the population over a vast area. But in dispersed actions under any circumstances, attention should be paid to the following: (1) no absolutely even dispersion of forces should be made, but a larger part of the forces should be kept at a place conveniently situated for its flexible employment so that, on the one hand, any possible exigency can be readily met and, on the other, the dispersed units can be used to fulfil the main task; and (2) the dispersed units should be assigned clearly defined tasks, fields of operation, specific time limits and rendezvous, and ways and means of liaison.

Concentration of forces, *i.e.* the method of "gathering parts into a whole," is adopted largely for the annihilation of an enemy on the offensive; it is sometimes adopted for the annihilation of certain stationary forces of the enemy when he is on the defensive. Concentration of forces does not mean absolute concentration, but the massing of the main forces in a certain important direction while retaining or dispatching a part of the forces in other directions for purposes of containing, harassing or disrupting the enemy, or for work among the population.

Although flexible dispersion or concentration of forces in accordance with circumstances is the principal method in guerrilla warfare, we must also know how to shift (or transfer) our forces flexibly. When the enemy feels seriously threatened by the guerrillas he will send troops to suppress or attack them. Hence guerrilla units should ponder over the circumstances they are in: if it is possible for them to fight, they should fight right on the spot; if not, they should not miss the opportunity to shift themselves swiftly to some other direction. Sometimes the guerrillas, for the purpose of smashing the enemy units separately, may, after annihilating an enemy force in one place, shift themselves immediately to another direction to wipe out another enemy force; sometimes the guerrillas, finding it inadvisable to fight in one place, may have to disengage themselves immediately from the enemy there and engage him in another direction. If the enemy's forces at a place are particularly strong, the guerrilla units should not stay there long, but should shift their positions as speedily as a torrent or a whirlwind. In general, the shifting of forces should be done secretly and swiftly. Ingenious devices such as making a noise in the east while attacking in the west, appearing now in the south and now in the north, hit-and-run and night action should be constantly employed to mislead, entice and confuse the enemy. . . .

—ANNE FREEMANTLE
(translator)

ANDRE MALRAUX:

FROM

Anti-Memoirs

*André Malraux was born in Paris in 1901 and, after study-
ing art, took part in an archaeological expedition to Indo-
china in 1923. He was involved in anticolonial activities
there and later worked with the Chinese revolutionary movement in Santon and Shanghai. On his way back he made
important archaeological discoveries in Arabia and later
organized an international air squadron in Spain for the
Republicans during the Civil War.*

*As a tank officer in the French Army, after the out-
break of World War II Malraux was twice captured but
escaped and finished the war in command of the Alsace
brigade of the Resistance. He joined de Gaulle and became
one of his closest associates. He was Minister of Information after the liberation and Minister of Culture after the
creation of the Fifth Republic. His earlier works are mainly
concerned with man in action, especially* The Human Condition *(1933), and* Days of Hope *(1937). His later works,
such as the* Voices of Silence *(1951), are concerned with
art and man in creation.*

Ninety thousand men, women and children would attempt to
break through the blockade, as Chu Teh had done in the Ching-kang
mountains. Little by little, the front-line army was replaced by partisans. On 16 October 1934, concentrated in southern Kiangsi, it took
the enemy fortifications by storm, and veered westward. The Long
March had begun.

Mules were loaded with machine-guns and sewing-machines.
Thousands of civilians accompanied the army. How many would

remain in the villages—or in the cemeteries? How many of the dis-
mantled machines carried on mule-back would be found again one
day, buried along the seven-thousand-mile route? The partisans with
their red-tasselled pikes and their hats made of leaves that shook like
feathers would hold out for a long time yet—some of them for three
years. The Nanking forces killed them, but Mao's army marched on.

In one month, harassed from the air, it fought nine battles,
broke through four lines of blockhouses and a hundred and ten regiments. It lost a third of its men, decided to keep only its military
equipment and a few field printing-presses, stopped advancing towards the north-west (which baffled the enemy but slowed its march
considerably). Chiang Kai-shek had gathered his forces behind the
Yangtze, and destroyed the bridges. But a hundred thousand men
and their artillery were awaiting Mao before the Kweichow River.
The Reds wiped out five divisions, held a meeting of their Central
Committee in the governor's palace, enrolled fifteen thousand
deserters, and organized youth cadres. But the "golden sands river"
of the poems had yet to be crossed. Mao turned southwards, and in
four days was fifteen miles from Yunnanfu, where Chiang Kai-shek
had established himself. It was a diversion, for the main body of the
Red Army was marching northwards to cross the river there.

It was the Tatu River, no less difficult to cross than the Yangtze,
and where the last army of the Taipings had been wiped out by the
imperial forces. Moreover it could only be reached through the vast
forest of the Lolos, where no Chinese army had ever penetrated. But
a few Red officers who had served in Szechwan had once set free
some Lolo chiefs, and Mao negotiated with these unsubdued tribes as
he had done with all the villages his soldiers had passed through.
"The government army is the common enemy." To which the tribes
responded by asking for arms, which Mao and Chu Teh ventured to
give them. The Lolos then guided the Reds through their forests
where the Nanking air force lost all trace of them—to the Tatu ferry,
which together they captured in a surprise attack.

It would have taken weeks for the army to cross the river by
means of this ferry. Chiang Kai-shek's airmen, reconnoitring the
river, had found the columns again. His armies had by-passed the
forest, and would soon be ready to give battle once more. This was
the time when Nanking spoke of the funeral march of the Red Army.

There was only one bridge, much further up the river, between
steep cliffs across a rushing torrent. The army, exposed to continuous

bombardment, advanced by forced marches through a storm along a narrow trail above the river, which by night reflected the thousands of torches tied to the soldiers' backs. When the advance guard reached the bridge, it found that half the wooden flooring had been burnt down.

Facing them, on the opposite bank, the enemy machine-guns.

All China knows the fabulous gorges of its great rivers, the fury of the waters pent up by sheer peaks which pierce the heavy, low clouds under the echoing cries of the birds of prey. It has never ceased to picture this army of torches in the night, the flames of the dead sacrificed to the gods of the river; and the colossal chains stretching across the void, like those of the gates of Hell. For the bridge of Luting now consisted of the nine chains which supported its floor of planks and two chains on either side which served as handrails. With the wooden roadway burnt, there remained these thirteen nightmare chains, no longer a bridge but its skeleton, thrusting over the savage roar of the waters. Binoculars revealed the intact section of the roadway and a voluted pavilion from behind which came the crackle of machine-gun fire.

The Red machine-guns opened up. Under the whistling hail of bullets, volunteers dangling from the freezing chains began to advance, link after enormous link—white caps and white cross-belts standing out in the mist—swinging their bodies to heave themselves forward. One after another they dropped into the raging waters, but the lines of dangling men, swaying from their own efforts and the force of the wind whistling through the gorges, advanced inexorably towards the opposite bank. The machine-guns easily picked off those who were clinging to the four supporting chains, but the curve of the other nine chains protected the men advancing below them, grenades at their belts.

It is the most celebrated legend of Red China. In the communist store in Hong-Kong I had seen, first of all, the exodus, strung out for mile upon mile; the peasant army preceding the civilians bent double under their burdens like rows of men hauling barges; a multitude as bowed as that which accompanied the Partition of India, but resolutely prepared for battles unknown. Three thousand miles they travelled, liberating village after village on the way, for a few days or a few years; and here were those stooping bodies which seemed to have risen from the tomb of China, and beyond the gorges, those chains stretching across History. Everywhere, chains belong to the darker side of man's imagination. They used to be part of the equipment of dungeons—and still were, in China, not so long ago—and their outline seems the very ideogram of slavery. Those hapless men with one arm hanging limp under the bullets are still watched by the indigent masses of China as the other hand opens above the roar of an ageless gulf. Other men followed them, whose hands did not open. In the memory of every Chinese, that string of dangling men swaying towards freedom seem to be brandishing aloft the chains to which they cling. . . .

This famous episode nevertheless cost the Red Army fewer men than those which followed. It reached a region where the blockhouses of Nanking were still few, and regained the military initiative. But it still had to cross the high snow-covered passes of the Chiachin Mountains. It had been warm in June in the Chinese lowlands, but it was cold at fifteen thousand feet, and the cotton-clad men of the South began to die. There were no paths; the army had to build its own track. One army corps lost two thirds of its animals. Mountain upon mountain, soon corpse upon corpse: one can follow the Long March by the skeletons fallen under their empty sacks; and those who fell for ever before the peak of the Feather of Dreams, and those who skirted the Great Drum (for the Chinese, the drum is the bronze drum) with its vertical faces in the endless jagged immensity of the mountains. The murderous clouds hid the gods of the Tibetan snows. At last the army with the moustaches of hoar-frost reached the fields of Maokung. Down below, it was still summer. . . .

There were 45,000 men left.

The Fourth Army and the vague soviet authorities of Sungpan awaited Mao there. The Red forces now mustered a hundred thousand soldiers; but after a disagreement which allowed Nanking a successful offensive, Mao set off again towards the Great Grasslands with thirty thousand men. Chu Teh stayed behind in Szechwan.

The Great Grasslands also meant dense forest, the sources of ten great rivers, and above all the Great Swamplands, occupied by autonomous tribes. The queen of the Man-tze tribe gave orders that anyone who made contact with the Chinese, Red or otherwise, was to be boiled alive. For once, Mao failed to negotiate. Empty dwellings, vanished cattle, narrow defiles in which the tribesmen rolled boulders down on them. "A sheep costs a man's life." There remained fields of green corn, and giant turnips each of which Mao said, would feed fifteen men. And the Great Swamplands.

The army advanced, guided by native prisoners. Anyone who left the trail vanished. Endless rain in the immensity of the sodden grasslands and stagnant waters, under the white mists or the livid sky. . . .

On 20 October 1935, at the foot of the Great Wall, Mao's soldiers, wearing hats of leaves and mounted on little shaggy ponies like those of the pre-historic cave-paintings, joined up with the three communist armies of Shensi, of which Mao took command. He had only twenty thousand men left, of whom seven thousand had been with him all the way from the South. They had covered six and a half thousand miles. Almost all the women had died, and the children had been left along the way.

The Long March was at an end.

—TERENCE KILMARTIN
(translator)

BASIL LIDDELL HART:

FROM

The Real War;

FROM

Strategy

Basil Liddell Hart was born in Paris in 1895, the son of an English clergyman. After attending Cambridge he joined the army as an infantry officer in 1914. After being gassed on the Western Front, he began to develop infantry training methods and to engage in tactical studies. Invalided out of the army, he became Military Correspondent of the Lon-don Daily Telegraph *and then of the* Times *and a prolific writer of books on war. For a time he was adviser to the war minister, securing important military reforms, but his opposition to some of Churchill's war policies led to his eclipse in Britain—although his ideas remained influential in other countries, especially Germany. In later years he turned increasingly to military history, lectured at American universities and colleges, and was widely recognized as a pioneer of strategy. He died in 1970.*

FROM *The Real War*

The third battle of Ypres, which came to be known as Pass-chendaele, was launched on the Western Front in 1917, by a reluctant decision of Lloyd George—under pressure from Haig and his supporters. It cost 250,000 casualties and many reputations.

For the ominous condition of the French army, the crisis at sea caused by the submarine campaign, and the need to second the still possible Russian offensive, combined to justify Haig's decision in May, the situation had radically changed before the main offensive was actually launched on July 31st. In war all turns on the time factor. By July the French army, under Pétain's treatment, was recuperating, if still convalescent, the height of the submarine crisis was past, and the revolutionary paralysis of the Russian army was clear. Nevertheless, the plans of the British High Command were unchanged.

The historian may consider that insufficient attention was given to the lessons of history and of recent experience, and of material facts in deciding both upon the principle of a major offensive and upon its site. The axis of the attack diverged from instead of converging on, the German main communication, so that an advance could not vitally endanger the security of the enemy's position in France. Haig, curiously, was to adopt here the same eccentric direction of advance which a year later his advice prevented Foch and Pershing from taking on the other flank of the Western Front.

Thus an advance on the Belgian coast offered no wide strategic results, and for the same reason it was hardly the best direction, even as a means of pinning and wearing down the enemy's strength on a profitable basis. Moreover, the idea that Britain's salvation from starvation depended on the capture of the submarine bases on this coast had long since been exploded, for the main submarine campaign was being conducted from German ports. In fairness, however, one should add that this mistaken belief was impressed on Haig by the Admiralty.

Worse still, the Ypres offensive was doomed before it began—by its own destruction of the intricate drainage system in this part of Flanders. The legend has been fostered that these ill-famed "swamps of Passchendaele" were a piece of ill-luck due to the heavy rain, a natural and therefore unavoidable hindrance that could not be foreseen. In reality, before the battle began, a memorandum was sent by Tank Corps Headquarters to General Headquarters pointing out that if the Ypres area and its drainage were destroyed by bombardment, the battlefield would become a swamp.

. . . Nearly two months passed before the preparations for the main advance were completed, and during that interval the Germans had ample warning to prepare counter-measures. These comprised a new method of defence, as suited to the waterlogged ground as the

British offensive methods were unsuited. Instead of the old linear system of trenches they developed a system of disconnected strong points and concrete pill-boxes, distributed in great depth, whereby the ground was held as much as possible by machine-gun and as little as possible by men.

While the forward positions were lightly occupied, the reserves thus saved were concentrated in rear for prompt counter-attack, to eject the British troops from the positions they had arduously gained. And the further the British advanced the more highly developed, naturally, did they find the system. Moreover, by the introduction of mustard gas the Germans scored a further trick, interfering seriously with the British artillery and concentration areas. . . .

On July 22nd, the bombardment began, by 2300 guns, to continue for ten days, until at 3:50 A.M. on July 31st the infantry advanced on a fifteen-mile front to the accompaniment of torrential rain.

Thus, when on November 4th, a sudden advance by the 1st Division and 2nd Canadian Division gained the empty satisfaction of occupying the site of Passchendaele village, the official curtain was at last rung down on the pitiful tragedy of "Third Ypres." It was the long overdue close of a campaign which had brought the British armies to the verge of exhaustion, one in which had been enacted the most dolorous scenes in British military history, and for which the only justification evoked the reply that, in order to absorb the enemy's attention and forces, Haig chose the spot most difficult for himself and least vital to the enemy. Intending to absorb the enemy's reserves, his own were absorbed.

Perhaps the most damning comment on the plan which plunged the British Army in this bath of mud and blood is contained in an incidental revelation of the remorse of one who was largely responsible for it. This highly-placed officer from General Headquarters was on his first visit to the battle front—at the end of the four months' battle. Growing increasingly uneasy as the car approached the swamp-like edges of the battle area, he eventually burst into tears, crying "Good God, did we really send men to fight in that?" To which his companion replied that the ground was far worse ahead. If the exclamation was a credit to his heart it revealed on what a foundation of delusion and inexcusable ignorance his indomitable "offensiveness" had been based.

FROM *Strategy*

> *The original outline of the strategy of indirect approach was published in the author's* Decisive Wars in History *in 1929 and elaborated in editions of* The Strategy of Indirect Approach.

The perfection of strategy would be, therefore, to produce a decision without any serious fighting. History, as we have seen, provides examples where strategy, helped by favourable conditions, has virtually produced such a result—among the examples being Caesar's Ilerda campaign, Cromwell's Preston campaign, Napoleon's Ulm campaign, Moltke's encirclement of MacMahon's army at Sedan in 1870, and Allenby's 1918 encirclement of the Turks in the hills of Samaria. The most striking and catastrophic of recent examples was the way that, in 1940, the Germans cut off and trapped the Allies' left wing in Belgium, following Guderian's surprise breakthrough in the centre at Sedan, and thereby ensured the general collapse of the Allied armies on the Continent.

While these were cases where the destruction of the enemy's armed forces was economically achieved through their disarming by surrender, such "destruction" may not be essential for a decision, and for the fulfilment of the war-aim. In the case of a state that is seeking, not conquest, but the maintenance of its security, the aim is fulfilled if the threat be removed—if the enemy is led to abandon his purpose.

The defeat which Belisarius incurred at Sura through giving rein to his troops' desire for a "decisive victory"—after the Persians had already given up their attempted invasion of Syria—was a clear example of unnecessary effort and risk. By contrast, the way that he defeated their more dangerous later invasion and cleared them out of Syria, is perhaps the most striking example on record of achieving a decision—in the real sense, of fulfilling the national object—by pure strategy. For in this case, the psychological action was so effective that the enemy surrendered his purpose without any physical action at all being required.

While such bloodless victories have been exceptional, their rarity enhances rather than detracts from their value—as an indication of latent potentialities, in strategy and grand strategy. Despite many centuries' experience of war, we have hardly begun to explore the field of psychological warfare.

It rests normally with the government, responsible for the grand strategy of a war, to decide whether strategy should make its contribution by achieving a military decision or otherwise. Just as the military means is only one of the means to the end of grand strategy—one of the instruments in the surgeon's case—so battle is only one of the means to the end of strategy. If the conditions are suitable, it is usually the quickest in effect, but if the conditions are unfavourable it is folly to use it.

Let us assume that a strategist is empowered to seek a military decision. His responsibility is to seek it under the most advantageous circumstances in order to produce the most profitable result. Hence *his true aim is not so much to seek battle as to seek a strategic situation so advantageous that if it does not of itself produce the decision, its continuation by a battle is sure to achieve this.* In other words, dislocation is the aim of strategy; its sequel may be either the enemy's dissolution or his easier disruption in battle. Dissolution may involve some partial measure of fighting, but this has not the character of a battle.

* * *

How is the strategic dislocation produced? In the physical, or "logistical," sphere it is the result of a move which (*a*) upsets the enemy's dispositions and, by compelling a sudden "change of front," dislocates the distribution and organization of his forces; (*b*) separates his forces; (*c*) endangers his supplies; (*d*) menaces the route or routes by which he could retreat in case of need and reestablish himself in his base or homeland.

A dislocation may be produced by one of these effects, but is more often the consequence of several. Differentiation, indeed, is difficult because a move directed towards the enemy's rear tends to combine these effects. Their respective influence, however, varies and has varied throughout history according to the size of armies and the complexity of their organization.

In the psychological sphere, dislocation is the result of the impression on the commander's mind of the physical effects which we have listed. The impression is strongly accentuated if his realization of his being at a disadvantage is *sudden,* and if he feels that he is unable to counter the enemy's move. *Psychological dislocation fundamentally springs from this sense of being trapped.*

This is the reason why it has most frequently followed a physical move on the enemy's rear. An army, like a man, cannot properly defend its back from a blow without turning round to use its arms in the new direction. "Turning" temporarily unbalances an army as it does a man, and with the former the period of instability is inevitably much longer. In consequence, the brain is much more sensitive to any menace to its back.

In contrast, to move directly on an opponent consolidates his balance, physical and psychological, and by consolidating it increases his resisting power. For in the case of an army it rolls the enemy back towards their reserves, supplies, and reinforcements, so that as the original front is driven back and worn thin, new layers are added to the back. At the most, it imposes a strain rather than producing a shock.

Thus a move round the enemy's front against his rear has the aim not only of avoiding resistance on its way but in its issue. In the profoundest sense, it takes the *line of least resistance.* The equivalent in the psychological sphere is the *line of least expectation.* They are the two faces of the same coin, and to appreciate this is to widen our understanding of strategy. For if we merely take what obviously appears the line of least resistance, its obviousness will appeal to the opponent also; and this line may no longer be that of least resistance.

In studying the physical aspect we must never lose sight of the psychological, and only when both are combined is the strategy truly an indirect approach, calculated to dislocate the opponent's balance.

The mere action of marching indirectly towards the enemy and on to the rear of his dispositions does not constitute a strategic indirect approach. Strategic art is not so simple. Such an approach may start by being indirect in relation to the enemy's front, but by the very directness of its progress towards his rear may allow him to change his dispositions, so that it soon becomes a direct approach to his new front.

Because of the risk that the enemy may achieve such a change of front, it is usually necessary for the dislocating move to be preceded by a move, or moves, which can best be defined by the term "distract" in its literal sense of "to draw asunder." The purpose of this "distraction" is to *deprive the enemy of his freedom of action,* and it should operate in both the physical and psychological spheres. In the physical, it should cause a distension of his forces or their diversion to unprofitable ends, so that they are too widely distributed, and too committed elsewhere, to have the power of interfering with one's own decisively intended move. In the psychological sphere, the same effect is sought by playing upon the fears of, and by deceiving, the opposing command.

Superior weight at the intended decisive point does not suffice unless that point cannot be reinforced *in time* by the opponent. It rarely suffices unless that point is not merely weaker numerically but has been weakened morally. Napoleon suffered some of his worst checks because he neglected this guarantee—and the need for distraction has grown with the delaying power of weapons.

A deeper truth to which Foch and other disciples of Clausewitz did not penetrate fully is that in war every problem, and every principle, is a duality. Like a coin, it has two faces. Hence the need for a well-calculated compromise as a means to reconciliation. This is the inevitable consequence of the fact that war is a two-party affair, so imposing the need that while hitting one must guard. Its corollary is that, in order to hit with effect, the enemy must be taken off his guard. Effective concentration can only be obtained when the opposing forces are dispersed; and, usually, in order to ensure this, one's own forces must be widely distributed. Thus, by an outward paradox, true concentration is the product of dispersion.

A further consequence of the two-party condition is that to ensure reaching an objective one should have *alternative objectives.* Herein lies a vital contrast to the single-minded nineteenth century doctrine of Foch and his fellows—a contrast of the practical to the theoretical. For if the enemy is certain as to your point of aim he has the best possible chance of guarding himself—and blunting your weapon. If, on the other hand, you take a line that threatens alternative objectives, you distract his mind and forces. This, moreover, is the most economic method of *distraction,* for it allows you to keep the largest proportion of your force available on your real line of operation—thus reconciling the greatest possible concentration with the necessity of dispersion.

The absence of an alternative is contrary to the very nature of war. It sins against the light which Bourcet shed in the eighteenth century by his most penetrating dictum that "every plan of campaign ought to have several branches and to have been so well thought out that one or other of the said branches cannot fail of success." This was the light that his military heir, the young Napoleon Bonaparte, followed in seeking always, as he said, to *"faire son thème en deux facons."* Seventy years later Sherman was to relearn the lesson from experience, by reflection, and to coin his famous maxim about "putting the enemy on the horns of a dilemma." In any problem where an opposing force exists, and cannot be regulated, one must foresee and provide for alternative courses. Adaptability is the law which governs survival in war as in life—war being but a concentrated form of the human struggle against environment.

BIBLIOGRAPHY OF SELECTIONS

A few of the source works for this collection are available only in other countries. Direct quotations are from specific translations and editions. For the convenience of the American reader, where possible, a list follows of editions of these works obtainable in the United States, although not necessarily chosen by the editors of this volume. Those which are no longer in print are available in many circulation and reference libraries.

In several cases, more than one edition of a work exists. No special criteria were used in selecting among them, and the reader may locate another more easily.

I. THE ANCIENT WORLD

From the Book of Judges: The King James Bible.

Thucydides. *The History of the Peloponnesian War*. Translated by Richard Crawley. New York: E. P. Dutton & Co., 1963.

Xenophon. *Hellenica and Anabasis*. 3 vols. Loeb Classical Library. Cambridge, Mass.: Harvard University Press.

Aristotle. *Politics*. Translated by Benjamin Jowett. New York: Modern Library, 1943.

Sun Tzu. *The Art of War*. Translated by Samuel B. Griffith. Oxford: The Clarendon Press, 1963.

Polybius. *The Histories of Polybius*. Translated from the text of F. Hultsch by Evelyn S. Shuckburgh. Bloomington: Indiana University Press, 1962.

Julius Caesar. *Caesar's Commentaries*. Translated by Jane Mitchell. Baltimore: Penguin Books, 1972.

Vergil. *Aeneid*. Translated by C. Day Lewis. New York: Doubleday & Co., 1952.

Onasander. *Military Essays* (Bound with *Military Essays*, Aeneas Tacticus; and *Military Essays*, Asclepiodotus). Loeb Classical Library: Cambridge, Mass.: Harvard University Press, 1923.

Tacitus. *The Annals of Tacitus.* Edited by F. R. Goodyear. 4 vols. New York: Cambridge University Press, 1972.

Arrian. *The Anabasis of Alexander, and Indica.* 2 vols. Loeb Classical Library. Cambridge, Mass.: Harvard University Press.

Vegetius. *The Military Institutions of the Romans.* Translated by John Clark. Edited by Thomas R. Phillips. Harrisburg, Pa.: Military Service Publishing Company, 1944.

Procopius. *History of the Wars.* 7 vols. Cambridge, Mass.: Harvard University Press.

Leo. *Tactica.* Extract from general history. Also published in United States in 1935 by Colonel Spaulding. (Rare.)

Juvaini. *The History of the World Conqueror.* Translated from the text of Mirza Muhammad Qazvini by John Andrew Boyle. 2 vols. Manchester: Manchester University Press, 1958.

II. RENAISSANCE AND REFORMATION

Niccolo Machiavelli. *The Prince.* Translated by W. K. Marriott. New York: E. P. Dutton & Co., 1908.

——. *The Art of War.* Revised ed. Translated by Ellis Farneworth. Indianapolis: Bobbs-Merrill, 1965.

Richard Hakluyt. *The Principal Navigations, Voyages, Traffiques and Discoveries of the English Nation, Made by Sea or Over-Land to the Remote and Farthest Distant Quarters of the Earth at Any Time Within the Compasses of These 1600 Yeeres.* 12 vols. New York: AMS Press, 1965.

Walter Raleigh. *The History of the World.* Edited by C. A. Patrides. Philadelphia: Temple University Press, 1971.

Hugo Grotius. *The Law of War and Peace.* Translated by Francis W. Kelsey with the collaboration of Arthur E. R. Boak et. al. Indianapolis: Bobbs-Merrill, 1962.

Thomas Hobbes. *Leviathan.* Edited by C. B. Macpherson. Baltimore: Penguin Books, 1968.

Oliver Cromwell. *Writings and Speeches of Oliver Cromwell.* Edited by William Cortez Abbott. Cambridge, Mass.: Harvard University Press, 1947.

George Savile Halifax. *Complete Works.* Edited by J. P. Kenyon. Baltimore: Penguin Books, 1969.

Daniel Defoe. *Memoirs of a Cavalier.* Edited by James J. Boulton. London: Oxford University Press, 1972.

III. THE AGE OF REASON

Jonathan Swift. *The Conduct of the Allies.* Edited by C. B. Wheeler. Oxford: The Clarendon Press, 1916.

Maurice de Saxe. *Reveries, or Memoirs Upon the Art of War.* Reprint of 1757 edition. Westport, Conn.: Greenwood Press.

Jean de Bourcet. *The Defence of Piedmont* by Spenser Wilkinson. London: Oxford University Press, 1927.

Frederick the Great. *Frederick the Great.* Translated by Jay Luvaas. New York: Free Press (Macmillan & Co.), 1966.

Henry Lloyd. *A Political and Military Rhapsody on the Invasion and Defence of Great Britain and Ireland.* London, 1779.

———. *The History of the Late War in Germany.* London, 1776 and 1782. 2 vols. (Rare copy in War Office Library in London).

James Wolfe. Letter to Major Rickson, 5 November 1757, from *Life and Letters of James Wolfe.* Edited by Beckles Wilson. London: William Heinemann Ltd., 1909.

Alexander Suvorov. *The Art of Victory.* New York, 1966.

Edward Gibbon. *Autobiography.* Edited by Lord Sheffield. London: Oxford University Press, 1962.

Jacques Antoine Guibert. *General Essay on Tactics.* 2 vols. Reprint of 1781 edition. Westport, Conn.: The Greenwood Press.

John Paul Jones. Letter to Admiral Kersaint, 1791, in *Selected Naval Records.* Oxford: The Clarendon Press. Also, *Memoirs of Rear-Admiral Paul Jones.* Reprint of 1830 edition. New York, Da Capo Press, Inc., 1972.

Robert Jackson. *A View of the Formation, Discipline, and Economy of Armies.* London: Parker, Furnivall, and Parker, 1845.

IV. THE REVOLUTION

Horatio Nelson. *The Trafalgar Memorandum.* On display at the British Museum. *The Life of Nelson: The Embodiement of the Sea Power of Great Britain* by A. T. Mahan. 2 vols. New York: Haskell House Publishers, Inc., 1969. Reprint of 1897 edition.

———. *Nelson's Last Diary, September 13–October 21, 1805.* Edited by Gilbert Hudson. London: E. Mathews, 1917.

Napoleon Bonaparte. *Napoleon's Maxims of War.* Translated by

Lieutenant-General Sir G. C. D'Aguilar. Kansas City, Mo.: Hudson-Kimberly Publishing Company, 1902.

——. *Napoleon's Memoirs.* Edited by Somerset de Chair. London: The Golden Cockerel Press, 1945.

Arthur W. Wellington. *Despatches, Correspondence and Memoranda.* 8 vols. Reprint of 1880 edition. Millwood, N.Y.: Kraus Reprint Company.

Private Wheeler. *Letters of Private Wheeler.* Edited by B. H. Liddell Hart. London: Michael Joseph, 1952.

Armand de Caulaincourt. *With Napoleon in Russia; the Memoirs of General de Caulaincourt.* From the original memoirs as edited by Jean Hansteau. Abridged and edited by George Libaire. New York: Grosset and Dunlap, 1935.

Henri Jomini. *The Art of War.* Translated by G. H. Mendell and W. P. Craighill. Westport, Conn.: Greenwood Press, 1971.

Karl von Clausewitz. *On War.* Translated by Anatol Rapoport. Baltimore: Penguin Books, 1969.

Stendhal (Marie-Henri Beyle). *The Charterhouse of Parma.* Translated by M. R. Shaw. Baltimore: Penguin Books, 1968.

V. THE LATER NINETEENTH CENTURY

Helmuth von Moltke. *The Military Works (Militarische Werke).* Edited by the Prussian General Staff, 1891–1894.

Abraham Lincoln. *Collected Works of Abraham Lincoln,* 9 vols. Edited by Roy P. Basler. New Brunswick, N.J.: Rutgers University Press, 1953.

Herman Melville. *Works.* New York: Russell and Russell, 1963. Volume 6, *White-Jacket.*

William Tecumseh Sherman. Letter to Major R. M. Sawyer, 31 January 1864 in *The Sherman Letters.* Edited by Rachel S. Thorndike (reprint of the 1894 edition). New York: AMS Press, 1972.

William Howard Russell. *Despatches from the Crimea 1854–1856.* Edited by Nicolas Bentley. New York: Hill and Wang, 1966.

Charles Ardant du Picq. *Battle Studies: Ancient and Modern Battle.* Translated by Colonel John N. Greely and Major Robert C. Cotton. New York: Macmillan & Co., 1921.

Leo Tolstoy. *War and Peace.* Translated by Constance Garnett. New York: Thomas Y. Crowell Company, 1976.

Ivan Stanislavovich Bloch. *Modern Weapons and Modern War* (abridgment of *War of the Future*). London: Grant Richards, 1900.

Alfred von Schlieffen. *The Great Memorandum*. Translated by Andrew and Eva Wilson; edited by Gerhard Ritter. London: Oswald Wolff, 1958.

Emory Upton. *The Military Policy of the United States*. Reprint of 1904 edition. Westport, Conn.: Greenwood Press.

Alfred Thayer Mahan. *The Influence of Sea-Power upon the French Revolution and Empire, 1793–1812*. 2 vols. Westport, Conn.: Greenwood Press, 1968.

Colmar Von der Goltz. *The Nation in Arms*. London: William Heinemann Ltd., 1887.

Hans Delbrück. *The Art of War*. Translated by Dr. Peter Paret in *Military Affairs* Magazine, 1966. (Originally published in Berlin under the title *Geschichte der Kriegskunst im Rahmen der politischen Geschichte,* 1800).

Ferdinand Foch. *The Principles of War*. Translated by J. De Morinni. Reprint of 1918 edition. New York: AMS Press, 1970.

George Francis Henderson. *Stonewall Jackson and the American Civil War*. Abridged ed. Gloucester, Mass.: P. Smith, 1968.

Julian S. Corbett. *Some Principles of Maritime Strategy*. New York: AMS Press, 1972.

VI. THE TWENTIETH CENTURY

Halford MacKinder. *The Geographical Pivot of History*. London: 1904. Reprinted by the Royal Geographical Society, London, 1969.

Jean Colin. *Transformations of War*. Translated by L. Pope-Hennessy. Reprint of 1912 edition. Westport, Conn.: Greenwood Press.

Erich Ludendorff. *The Nation at War*. Translated by A. S. Rappoport. London: Hutchinson and Company, 1936.

Herbert George Wells. *The War of the Air*. Reprinted by Penguin Books, 1941.

Giulio Douhet. *The Command of the Air*. Translated by Dino Ferrari. Reprint of 1942 edition. New York: Arno Press, 1972.

Lenin. *War and Socialism* in *Works of V. I. Lenin*. New York: International Publishers, 1929.

——. *Left-Wing Communism: An Infantile Disorder.* New York: International Publishers, 1940.

Stephen Crane. *The Red Badge of Courage.* New York: Modern Library, 1951.

Marcel Proust. *Remembrance of Things Past.* 2 vols. New York: Random House, 1934.

Winston Churchill. *The River War.* New York: Universal Publishing and Distributing Corporation, 1964.

John Frederick Charles Fuller. *Memoirs of an Unconventional Soldier.* London: I. Nicholson and Watson, 1936.

William Mitchell. *Winged Defense.* Port Washington, N.Y.: Kennikat Press, 1971.

Leon Trotsky. *The History of the Russian Revolution.* Translated by Max Eastman. Ann Arbor: University of Michigan Press, 1932.

Douglas MacArthur. *Report of the Chief of Staff* in *A Soldier Speaks: Public Papers and Speeches of General of the Army Douglas MacArthur.* New York: Praeger Publishers, 1965.

T. E. Lawrence. *Evolution of a Revolt: Early Postwar Writings of T. E. Lawrence.* Edited by Stanley and Rodelle Weintraub. University Park: Pennsylvania State University Press, 1968.

——. *Seven Pillars of Wisdom, a Triumph.* Garden City, N.Y.: Doubleday, 1966.

Adolf Hitler. *Mein Kampf.* Translated by Ralph Manheim. Boston: Houghton Mifflin, 1943.

Albert Speer. *Inside the Third Reich.* Translated by Richard and Clara Winston. New York: Macmillan, 1970.

Charles de Gaulle. *The Edge of the Sword.* Translated by Gerard Hopkins. New York: Criterion Books, 1960.

Mao Tse-tung. *On Guerrilla Warfare.* Translated by Samuel B. Griffith. New York: Praeger, 1961.

André Malraux. *Anti-Memoirs.* Translated by Terence Kilmartin. New York: Holt, Rinehart and Winston, 1968.

Basil Liddell Hart. *The Real War, 1914–1918.* Boston: Little, Brown, 1930.

——. *Strategy.* 2nd rev. ed. New York: Praeger, 1967.

Acknowledgments

The editors of *The Sword and the Pen* and the Thomas Y. Crowell Company wish to thank the following publishers, agents, and translators for granting permission to reprint selections from the following copyrighted material. All possible care has been taken to trace ownership of material included and to make full acknowledgment for its use.

THE ARMY QUARTERLY for *The Evolution of a Revolt* by T. E. Lawrence.

JONATHAN CAPE LTD. AND THE SEVEN PILLARS TRUST for *The Seven Pillars of Wisdom* by T. E. Lawrence.

CHATTO AND WINDUS LTD. AND THE ESTATE OF C. K. SCOTT-MONCRIEFF for *Remembrance of Things Past* by Marcel Proust, translated by C. K. Scott-Moncrieff.

COWARD, MC CANN & GEOGHEGAN, INC. for *The Command of the Air* by Giulio Douhet, translated by Dino Ferrari. Copyright 1942 by Coward, McCann & Geoghegan, Inc. Reprinted by permission.

DOUBLEDAY & COMPANY, INC. for *The Seven Pillars of Wisdom* by T. E. Lawrence. Copyright 1926, 1935 by Doubleday & Company, Inc. Reprinted by permission of Doubleday & Company, Inc.

FABER AND FABER LIMITED for *The Edge of the Sword* by Charles de Gaulle, translated by Gerard Hopkins. Reprinted by permission.

HAMISH HAMILTON for *Anti-Memoirs* by André Malraux.

THE HAMLYN GROUP LTD. for *The River War* by Winston Churchill. Reprinted by permission.

HODDER & STOUGHTON LTD. for *The Transformation of War* by J. Colin.

HOLT, RINEHART AND WINSTON for *Anti-Memoirs* by André Malraux. Translated by Terence Kilmartin. Copyright © 1968 by Holt, Rinehart and Winston and Hamish Hamilton Ltd. Reprinted by permission of Holt, Rinehart and Winston, Publishers.

HUTCHINSON PUBLISHING GROUP LTD. for *The Nation at War* by Erich Ludendorff.

MACMILLAN PUBLISHING COMPANY, INC. for *Inside the Third Reich* by Albert Speer. Copyright © 1970 by Macmillan Publishing Company, Inc. Reprinted by permission.

THE NEW AMERICAN LIBRARY, INC. AND ANNE FREEMANTLE, translator, for *On Guerrilla Warfare* and *Anti-Japanese Guerrilla Warfare* by Mao Tsetung.

RANDOM HOUSE INC. for *The Guermantes Way* by Marcel Proust. Copyright 1925 and renewed 1953 by Random House, Inc. Reprinted from *Remembrance of Things Past,* Volume I, by Marcel Proust, translated by C. K. Scott-Moncrieff, by permission of Random House, Inc.

MARGARITA ALISON STARR for *Memoirs of an Unconventional Soldier* by J. F. C. Fuller.

THE UNIVERSITY OF MICHIGAN for *The Russian Revolution* by Leon Trotsky, translated by Max Eastman. Copyright by the University of Michigan 1932, 1933, 1960. Renewed 1961.

A. P. WATT & SON for *The War in the Air* by H. G. Wells. Reprinted by permission of the Estate of H. G. Wells.

GEORGE WEIDENFELD & NICOLSON LTD. for *Inside the Third Reich* by Albert Speer.